The Chinese
Information War

The Chinese Information War

Espionage, Cyberwar, Communications Control and Related Threats to United States Interests

SECOND EDITION

Dennis F. Poindexter

McFarland & Company, Inc., Publishers

Jefferson, North Carolina

LIBRARY OF CONGRESS CATALOGUING-IN-PUBLICATION DATA

Names: Poindexter, Dennis F., 1945– author.
Title: The Chinese information war : espionage, cyberwar, communications control and related threats to United States interests / Dennis F. Poindexter.
Description: Second edition. | Jefferson, North Carolina : McFarland & Company, Inc., Publishers, 2018. | Includes bibliographical references and index.
Identifiers: LCCN 2018024097 | ISBN 9781476672717 (softcover : acid free paper) ∞
Subjects: LCSH: Cyberspace operations (Military science) | Information warfare—China. | China—Relations—United States. | United States—Relations—China. | Information warfare—United States. | National security—United States. | Computer security—United States. | Espionage, Chinese. | China—Strategic aspects.
Classification: LCC U167.5.C92 P645 2018 | DDC 355.4—dc23
LC record available at https://lccn.loc.gov/2018024097

BRITISH LIBRARY CATALOGUING DATA ARE AVAILABLE

ISBN (print) 978-1-4766-7271-7
ISBN (ebook) 978-1-4766-3176-9

Front cover images © 2018 iStockphoto

Printed in the United States of America

McFarland & Company, Inc., Publishers
Box 611, Jefferson, North Carolina 28640
www.mcfarlandpub.com

Table of Contents

Preface

This book is about a war that many will doubt we are in: an information war with China. To understand this is to believe that we do not have war anymore, at least in the traditional sense of it. The new type of war is China's bending a country's will to its own. It is clever, broadly applied, successful, and aimed directly at the United States.

The 1990s were the years of information warfare, at least judging by the two-inch-thick treatise *Information Warfare: Legal, Regulatory and Organizational Considerations for Assurance*, published by the Joint Staff of the Pentagon in 1996; but it was being talked about long before it became fashionable. The Chinese were following the strategies before that, but they may have been encouraged by our concentration on it. They have applied themselves to it and have done well.

This war is neither conventional nor accidental. The U.S. military expanded the doctrine and made it public, but they seldom used it themselves. Our military is at a disadvantage in its application because it is part of a system of government that is democratic and decentralized, and that separates the government from commercial business. This is a system that has served us well, but it is not one that China sees as a path to the top of the world's food chain.

This is not a "how to" book of strategies that might be developed to fight a war. It is a way to organize what the Chinese are already doing, to make sense of it. Until we see ourselves as being in a war, we cannot begin to fight it effectively. It doesn't fit our idea of war because the Chinese have changed the nature of war to carry it out. The First Principle of War is never to be at war with anyone. This is the Escher drawing of war—a stairway that never goes where it should. They will never admit that what they are doing could be interpreted as war. Denial is a large part of this form of war.

We tend to think of war as something fought by the military, when we know the militaries of the world are often augmented by other types of covert actions, usually carried out by the intelligence communities of different coun-

1

tries. We know that a military attack force can have ground troops who are not in uniform and not working for a military commander. There are helicopters overhead that are "different" from some of the others and are not manned by pilots working for a military. There are support functions that give maps and warnings to attacking military forces. Some of the forces operate from front companies that look like legitimate businesses but are really covered operations. The Chinese use their companies as covers for their government and military, but they are not the only country that operates them. If we don't see them for what they are, they are doing their operational security the way it is supposed to be done. Information war should be below the enemy's radar.

The gap that this book covers is an area of information warfare that is called "black" and involves classified national security information, which is not supposed to be written about in the public media. The irony, of course, is that people do actually write about things that are black, and the government even approves some of the things that are said. We talk about things we do in war, yet we hold these things to be among our most valued secrets.

In this book I've done the best I can to describe some of the aspects of this type of warfare without discussing things that could be related to United States military or intelligence community capabilities. Sometimes that means being vague about what the U.S. might be able to do, or how an enemy might be able to develop other capabilities. Where possible, I have used open sources and what hackers are doing, even though those groups may not be sponsored by a government agency.

Hackers have caught up to what the governments were doing ten years ago, and the concern today is that terrorists will do the same. It is probably inevitable that this will happen, and when it does, there will be almost no deterrent that can stop a stateless group from causing us damage that will be painful. I have tried not to encourage them or point them in any particular direction. Too many people speculate on how to hurt us, without thinking about how that might benefit someone who hasn't thought about the subject very much.

I wrote this book for a general audience and not just for military people. The military does not pay much attention to doctrine unless it helps them win wars. This is not a war that the military would fight alone. My purpose is to educate by attempting to teach concepts, not techniques, of warfare. If my audience believes we might already be at war with China, the book is a success.

There are several congressional reports that are very authoritative and well written, and, for the most part, I have used these as sources. They are the only unclassified sources for some of this material. The U.S.–China Economic and Security Review Commission has provided the best sources of any committee of Congress and has done so more frequently than most others. They are both authoritative and thorough.

I have said what I think China has been doing, and in this edition included a few things about Russia, North Korea, and Iran without mentioning some of the U.S. capabilities to counterattack. Some of my critics think China is weak and does much of their information war out of fear of the rest of the world. I doubt that. They could not take on countries like the United States if they believed they were too weak to do so. Their strategy is to wait for a time when the advantage is theirs. At the same time, they know their own strength.

I believe the Russians and Chinese work together more than we currently know, in information warfare. Their techniques and strategies are too much alike for it to be coincidental.

We are behind the Chinese in areas that are going to cause us great harm if we don't start catching up. But it is not something you will hear discussed in the halls of Congress or the Pentagon. Exactly how we are behind is one of our best kept secrets and will stay that way. We have to learn to fight in similar ways, and we have to do it quietly.

This book could not have been written without the support and inspiration of my wife, Virginia. She is very smart and keeps good counsel.

1

No Wars Here

The First Principle of War
To throw by strategic movements the mass of an army, successfully, upon the decisive points of a theater of war, and
also upon the communications of the enemy, as much as possible, without compromising one's own.
—Major General A. H. Jomini,
The Art of War, Paris, 1838

In 1862, during the United States Civil War, the area around Antietam Creek in Sharpsburg, Maryland, saw a battle where soldiers walked relentlessly across an open field of fire while their comrades fell beside them. Soldiers threw themselves across a narrow bridge directly under fire from hills above as waves of them died, one after another. Quite a bit has changed since that war, but we sometimes need to walk in those fields to experience what soldiers must have felt while they fought. Thousands of tourists visit our region every year to do just that.

I walked with a few of them down a long, sloping hill at the battlefield, where four thousand troops were killed. It was a hot day; the sun was blazing. Every step was agonizing. I wondered what it must have been like to look at the fence row ahead and see those little puffs of smoke popping up almost everywhere, watching people drop on either side. The grass would have been wet with the blood of soldiers. It made the hair stand up on the back of my neck. It seemed like certain death.

You have to ask yourself how it is possible that soldiers believe this is part of their job—to die with near certainty—because someone thinks it is a sound military strategy. A soldier who looks down at that fence line has to know the truth. Yet years after Antietam, we got even better at killing soldiers without getting better at war.

In the fields of Verdun, France, soldiers were blasted by new weapons from the air and ground. Airplanes and machine guns made killing much easier and more efficient. The Germans had almost complete air superiority,

5

which they used to their advantage. The French had the fierceness of soldiers fighting for their home, and little else. From February to December of 1916, fighting surged and stopped with seemingly endless deaths among the combatants. When the battle was over, the Germans had lost 140,000 men and the French 162,000.

All the while, those soldiers, and the ones who have come after them, must have wished there was a better way to settle a dispute.

Now, there is.

Today, we can win territory, capture populations, influence our enemies, and engage our allies without having a war by any current definition. Soldiers enforce checkpoints inside another country, arrest political opponents, seize radio and television stations, take over public utilities, and declare the territory theirs. Is that war? Apparently not, since no government, press outlet or public view says it is. The occasional government official has called it *annexation*. Annexation is "the act of attaching, adding, joining, or uniting one thing to another. It is usually applied to land or fixtures as: the acquisition of land or territory by a nation, state, or municipality."[1] It is not war.

Adolf Hitler used the term frequently to describe his taking of various parts of countries around Germany just prior to World War II. Trying to stop the war Hitler threatened, the major powers of Europe conceded land owned by other countries. It seemed like a good way to stop a war from happening, but we all know how that strategy turned out.

Russia in Crimea and Ukraine, China in the South and East China Sea, and ISIS in the Middle East are not at war with their neighbors by almost any definition we find today, yet they have each seized territory others claimed and used some force to do it. Russia and China have strengthened their positions by convincing allies and enemies alike that these are peaceful takeovers of territory each owns. Governments and their political allies, even parts of the academic community, define war in such a way that the methods used in these circumstances fall outside the bounds. But if it cannot be war, what is it? *Annexation* describes what has occurred, yet that is not an acceptable term here any more than it was in the time of Adolf Hitler.

In 2012, Leon Panetta, who was Secretary of Defense at the time, was asked if we were at war with China, and he said, "I guess it depends on your definition of war." Some researchers want a formal definition that is recognized by scholars. There are many, but this one is recognized in law:

> A contest by force between two or more nations, carried on for any purpose, or armed conflict of sovereign powers or declared and open hostilities, or the state of nations among whom there is an interruption of pacific relations, and a general contention by force, authorized by the sovereign.... War does not exist merely because of an armed

attack by the military forces of another nation until it is a condition recognized or accepted by political authority of government which is attacked, either through an actual definition of war or other acts demonstrating such position.[2]

By this definition, the United States and China are not at war, and probably never will be. Mao Tse-tung, China's leader from 1949–1976, considered war something entirely different, "*War is* the highest form of struggle for resolving contradictions, when they have developed to a certain stage, between classes, nations, states, or political groups, and it has existed ever since the emergence of private property and of classes. Unless you understand the actual circumstances of war, its nature and its relations to other things, you will not know the laws of war, or know how to direct war, or be able to win victory."[3]

Uppsala University's Department of Peace and Conflict Research has been studying conflict since 1971 and can split and define everything from armed conflict to deaths caused by factions in Mexico's drug wars.[4] This group breaks events into conflicts that result in deaths of some combatants and non-combatants. Uppsala makes several distinctions in conflict, with one being the difference between civil war and interstate war:

> Civil war is armed fighting between the government of a state and one or more opposition groups concerning the government and/or territory of the state. Civil wars are distinct from interstate wars, that is, wars between two or more states. The criterion that the government of a state is one of the warring sides is necessary to distinguish civil war from other forms of organized violence within states, occurring between non-state actors.[5]

But what these definitions lack is an accounting for governments' covert actions to disguise who is actually involved in the conflict. Are those Russian-speaking soldiers really Ukrainian nationalists, or Russians who are "on leave" from their duties in the Federation? We have no way of knowing until we capture one, and even then we may have doubts. If they keep their mouths shut, their covert status is known only to them.

By formal definitions the Russian invasion of Crimea is not an armed conflict, and it was certainly not war because there were almost no deaths. By Uppsala's definition, Ukraine was in a civil war, as were Israel and Palestine in 2015. But in no case, using its definition, would we describe as war either the events in Crimea or the seizure of territory in the South China Sea. The logic that allows this to be true is clear enough, but it does not account for the kind of wars we have today. The nice thing about the new wars is the possibility of winning without casualties, or firing a shot. The soldiers at Antietam would have liked that part.

On the ground, *annexation* might not characterize what is happening to the lives of people caught up in it. In places such as Syria, where there is considerable fighting with tanks, airplanes, artillery, chemical weapons and small arms, we are said to have a civil war. If during that civil war, the Syrian government uses chemical weapons to kill children, and the United States fires 60 Tomahawk missiles at an air base in Shayrat, the actions on both sides are short of war.

Through all of this, we still call what is happening in Syria a civil war. The fact that the parties are fighting in Syria, Iraq, Afghanistan, Libya, Yemen, and in the border areas of Turkey seems to have not affected the name that was given to the conflict. Nations such as China, Russia, the United States, Turkey, Iraq, Iran and others fight with armed groups that are allied with nobody in particular. ISIS seized territory by force and held it the same way. They kept it through intimidation that included beheading people who did not agree with them. Governments cooperate in the fiction that there is no *interstate war* anywhere in the Middle East, while refugees flood into Europe. Those refugees are from more countries than Syria, although the majority seeking refuge are from there; but an equal number come from Afghanistan, Iraq, Kosovo, Albania, and Pakistan.[6] If we are looking for wars, we could look for refugees—they know where the wars really are.

Figure 5. U.S.S. *Ross* fires Tomahawk missile (Defense Department).

The missile strike in Shayrat was a response very few expected, but probably nobody was more surprised than Chinese president Xi Jinping at President Trump's Mar-a-Lago estate in Florida. The trip down to Florida on *Air Force One* was a get-to-know-you excursion, mostly friendly, with press coverage galore. The topics on the table centered around the trade deficit and North Korea. One would think that it was at least unfortunate timing to have the two together when the missile strike was made and announced. While it was not a great moment for diplomacy with China, it may prove to have been a defining moment for North Korea.

The U.S. has said it will not allow North Korea to threaten the use of nuclear weapons, yet North Korea has been doing that for years while being supported by China. The only difference now is North Korea's credibility. The threat of nuclear destruction is less credible when a country does not have nuclear weapons. As it tests weapons and refines its delivery systems, those threats become more credible. The analogy between the air strike in Syria and the warning the U.S. gave to North Korea was not lost on anyone: The world says it agrees that chemical weapons are not to be used, and there will be consequences for doing so; the same is true of nuclear weapons, in case Syria was wondering. The significance of the message is seen beyond the borders in the Middle East.

Part of what we see is the appearance of governments cooperating with one another to keep peace, even when there is none. They control information about the actions of multiple parties so as to portray them as below the threshold of war, the so-called "peace in our time" of Munich. Governments support our ability to ignore the occurrence of war. It looks very much like the description of Winston Churchill in a speech before the House of Commons the month after a "peace agreement" with Hitler's Germany. He said:

> And do not suppose this is the end. This is only the beginning of the reckoning. This is only the first sip, the first foretaste of a bitter cup which will be proffered to us year by year unless by a supreme recovery of moral health and martial vigour, we arise again and take our stand for freedom as in olden time.[7]

Hitler's invasion of Czechoslovakia using superior military force changed many contrary opinions, but not all. Patience was not one of Hitler's strong suits, a characteristic shared by the Russians. China will wait until the time is right, and for the rest of the world, it will be too late. The consequences may be far greater than we can imagine.

After World War II, we called the threat-of-war-without-war a Cold War. We have believed for most of that time that wars were hotly contested with destructive weapons that blew people up or made their lives miserable in other ways, but the Cold War was something different. At least, that is what we said. But from the standpoint of someone who lived through it, what seemed like peaceful tension between Russia, its proxies, and the United States and its proxies was only what the public saw. The chance of a larger or more menacing threat to humanity was just around the corner from the usual activities of this Cold War, yet we hardly ever mentioned those. There were moments when the Cold War could feel a little warm to the touch.

Early on in my career, I sat in my military commander's truck out on the end of a runway located in South Florida. A klaxon had announced an alert to the pilots and they were off to their planes, firing up the engines and waiting. We were watching our unit's nuclear-loaded B-52 bombers maneuver into position. They are huge aircraft that can move with grace when they have to, and they were sliding into line like they were tied together. We were the Strategic Air Command, part of a nuclear triad of defenses against our enemies, which we took to be mostly Russia.

There was a burst of radio traffic saying, "This is not an exercise," which was the first authoritative thing either of us had heard about what was going on. Those were words nobody in our group wanted to hear. Our exercises were annoying, but not life-threatening, and this was not going to go well if the Russian satellites saw those bombers and decided to launch their missiles to get as many of them as they could before they took off. In those days, our policy was "mutually assured destruction," a kind of Armageddon that satisfied both sides. If we launched, they launched, leaving large swatches of our earth wrapped in fire. That idea did not appeal to anyone who knew what it meant. I checked the skies above us more than once for missile contrails, even knowing what that would mean to someone on the ground.

After a few minutes that seemed like hours, our systems were ordered to stand down. At the time, none of us knew why or how we got so close to war. The Russians may not have noticed, or they failed to respond to our actions. We did not know what happened in Moscow until many years later, and it was a classified state secret for many years after that.

We kept the real meaning of what we were doing that day from the average person on the street because it was not something we thought our citizens should be troubled about. The effects of nuclear weapons will keep a rational person up nights, so this was for their own good. The same can be said for certain aspects of information war.

Information war involves highly classified state secrets at the top of governments, never discussed outside of special rooms and computer systems designed to be secure from anyone who wants these kinds of secrets. The facts in these closed discussions are almost never discussed with the public.

Keeping them secret protects national security, but it also keeps the innocents from worrying.

We cannot live without computers, or so says my granddaughter. We would be uncomfortable without electricity. We can't begin to understand the effects of a nuclear blast in the air over one of our major cities. Modern war contemplates both of those. Yet, unusual for such secrets, they are discussed openly in newspapers and books. Anyone can read the never-published U.S. Top Secret Presidential Policy Directive 20, which describes the operational considerations for launching offensive cyber activities. It is not just that we don't want our populations concerned about these kinds of things. Those effects are really disturbing for them when they are.

In 2016, General James Cartwright was convicted for disclosing some of those secrets to the *New York Times*, when his defense was that he tried to prevent the disclosure of facts. The creature of science that prompted his prosecution was Stuxnet, a computer worm that was constructed to destroy centrifuges used in making nuclear material in Iran.[8] The idea that the Iranian nuclear program could be delayed by such action may have been optimistic, but at least for some time it did what it was supposed to do.

The details of how the Stuxnet was developed, how it worked, how it was deployed, and the countries involved appeared in books and newspapers, yet the Stuxnet program was Top Secret in the Pentagon and the White House. This is what I have called the *dichotomy of secrets*, i.e., the seeming contradiction that allows sensitive state secrets to be published in places such as public newspapers. The Top Secret classification means that the disclosure of that data to an unauthorized person will cause "grave damage" to the United States. Whoever gives this type of data to a newspaper is committing a serious crime and is doing grave damage to the United States if it is published, yet this is done repeatedly.

Cartwright's lawyers said he tried to prevent these disclosures by heading off the publication of the information given to the *Times* by someone else. If that were true, the case likely would never have gone to trial. Either the Justice Department or the attorneys for General Cartwright were not telling the whole truth. We can only guess which one it might have been. The principle we are seeking lies in the illegal disclosure of facts to support a government's position, in this case that the Obama administration actually was doing something to try to stop Iran from developing a nuclear weapon.

At the time, that administration was developing an agreement with Iran and other countries for Iran to forgo nuclear weapons for a time. There was disagreement about whether such an agreement could be enforced or monitored. For the public to believe that the United States had the power to disrupt the Iranian program if that agreement did not work, someone had to disclose these secrets, and someone did. We just cannot be sure who that person was, and we will never find out. In one of President Obama's last official acts, General Cartwright was pardoned for the crime he may have never committed.

Fred Kaplan's book *Dark Territory*, a background of some of the events that produced our cyber policies, outlines some of the most recent cyber incidents that would be classified in any government circle because they disclose capabilities of the United States Intelligence and military communities in this cyberwar.[9] Most of what Kaplan writes about has classified aspects that cannot be discussed in public, yet his sources have allowed that to be done.

Curiously, Russia, North Korea and China have exposed many of these secrets by intelligence operations of their own. Intelligence services used to rely on the collection and analysis of intelligence information to advise their governments on what options to consider when reacting to events. They have always believed that it was better to not let the other country know what information was stolen, because doing so always leads to a dragnet for the source of that information. Sometimes that would mean the exposure of information about another country's activities—usually to discredit or expose an activity that could only be effective if kept secret. But in the last ten years, the intelligence services have started to believe that exposing secrets is one way to get the upper hand with different groups who read what is published. The director of the CIA, Mike Pompeo, alluded to this in April 2017 when he indicated that Wikileaks had sought people who could apply to the CIA presumably to seek secrets which would then be exposed:

> WikiLeaks walks like a hostile intelligence service and talks like a hostile intelligence service. It has encouraged its followers to find jobs at CIA in order to obtain intelligence. It directed Chelsea Manning in her theft of specific secret information. And it overwhelmingly focuses on the United States, while seeking support from anti-democratic countries and organizations.
>
> It is time to call out WikiLeaks for what it really is—a non-state hostile intelligence service often abetted by state actors like Russia. In January of this year, our Intelligence Community determined that Russian military intelligence—the GRU—had used WikiLeaks to release data of US victims that the GRU had obtained through cyber operations against the Democratic National Committee. And the report also found that Russia's primary propaganda outlet, RT, has actively collaborated with WikiLeaks.[10]

The accuracy of conclusions that follow publication of this kind of news is not of concern to intelligence services. They want argument, disruption of political processes, and weakness in any government they cannot control. They get the added benefit of discrediting some agencies in the United States and some of its allies. They are doing quite well with this approach, but this is only a very small part of information war called political warfare. We are now paying more attention to what the Russians do than to what the Chinese have been doing, and that may be a mistake, though an understandable one, in this case.

The Russians are accused of stealing internal secrets of the Democratic National Committee and publishing them to influence the United States 2016 elections. It will be many years before we know the truth of that assertion, but it makes for grand politics. The United States Intelligence Community gave briefings to congressional committees and the congressional representatives repeated some of the things they heard, adding their own interpretations. There were accusations that the theft of data was done to help Donald Trump win the election. There were implications that people in the Trump campaign may even have helped the Russians. Accusations and implications are not national security secrets, but their value lies in their ability to influence public opinion. If the accusation creates the thought that the president of the United States is not legitimately elected, it is successful. The accuracy of any of these statements is not germane to whether they can produce argument, distrust, or conflict in another government. Political parties act just like foreign intelligence services in publishing this kind of material.

Edward Snowden was part of this information war. He stole highly classified documents from the United States government and made them public. He escaped to China first, then went to Russia, where he still lives. That must tell us something about his inability to come home without some repercussions. His catalog of Top Secret information about U.S. intelligence collection operations, given freely to newspapers, goes on disrupting business, international relations and government operations of the United States. That is not something easily forgiven.

His documents are considerably more sensitive than the e-mail of the Democratic National Committee, yet the effects of their release are very much alike. The principle is to steal protected information and instead of keeping quiet about the content, expose it through the press, or use it internally to bolster economic and political advantage. The public can barely understand the ideas expressed in those documents, but governments understand them very well. The Snowden documents are blueprints for spying, i.e., for how to collect private information from public and private organizations and personal correspondence. It is not hacking for profit or hacking for personal information about a political opponent. This is state-sponsored activity to collect intelligence, but there is a new use for that data after it is collected. The ability to influence world events by stealing private communications, even highly classified state secrets, and making them public in mass, is a relatively new part of information war techniques, and its effectiveness may not yet be easily measured. So far, it seems to be successful.

There is a subtle difference between some of these thefts of data and what Director of National Intelligence James Clapper characterized in his December 2016 testimony to the Armed Services Committee. He said that an event such as the theft of security clearance records from the Office of Personnel Management—attributed to China—was different from the theft of information from something like a business, a kind of hacking China also does. The theft from OPM was an "act of espionage, when we and other nations do similar things." He said that acts of espionage do not lead to retaliation because we all live in glass houses. The U.S. and a number of other countries also run similar types of operations, and performing retaliatory strikes for this kind of collection is not done. From where he sits, that argument makes perfect sense, but from the view of those whose records were stolen, we might wonder why there is such a difference. Does it really matter to the victim if the information is made public or is just held by China? It is a long way from the personal impact of data thefts, experienced by every person with a current security clearance, to what this war is really about.

Russia and China are expansionist countries with appetites for territory. Russia declares a "New Russia," which is variously defined as a groups of territories that includes the southern portion of Ukraine, westward to Moldova, or in the context used by Vladimir Putin, the territories of the Czars of the early 1800s.[11] The latter stretches all the way to the Black Sea.

China has the "Nine-Dash Line" which, on a map, is literally nine large dashes starting at the northern tip of the Philippines, then running down to Brunei and back up the east coast of Vietnam. We see this in some context almost every week. It first appeared on Chinese maps in 1947, before the Communist revolution. It has been variously interpreted to mean a claim to territory, navigation rights, or fishing rights. It would appear to be easy to just ask China what the map means, but there seems to be no reason to clarify what it is intended to portray; perhaps we should never ask a question if we do not want to hear the answer.

Both Russia and China have made claims to territory that make Hitler's Germany seem not very ambitious, and at least the equivalent of the Japanese Empire. We can quibble about the ability of the two countries to achieve their objectives, but what we cannot quibble about is that they are trying, and succeeding.

Russia in Crimea and China in the South China Sea are annexing territory the same way Germany did before World War II, and no other country seems willing or able to stop them. But they are doing much more. Together with their allies, they are undermining democratic countries such as the U.S., the NATO allies, and the economies that make those countries viable. That strategy will eventually lead us to real war when the economic and political realities of what they have accomplished start to bite into the world's political alliances. To some extent, that is already happening in Europe, Southeast Asia, and the Middle East.

The second thing they both do is preserve their own political systems, even though they are dramatically different. China wants to preserve the Communist Party and its leadership in positions of power. A 2017 Defense Department report to Congress on China describes their approach as preserving more than just the Party. China's goals are to

- perpetuate CCP rule;
- maintain domestic stability;
- sustain economic growth and development;
- defend national sovereignty and territorial integrity;
- secure China's status as a great power and, ultimately, reacquire regional preeminence; and
- safeguard China's interests abroad.[12]

Russia is not communist anymore and seeks to keep the powerful oligarchs in power. Both China and Russia use information warfare on their own populations to keep them in line with those objectives.

We seem to believe this strategy of making war is not war at all. China has convinced us there is no war between its allies and ours. It uses the principles of information war in a meaningful way to combine the resources of a state and its allies against other states. It undermines the economic and political institutions of its enemies, and it is successful at it. But most of all it has allies, particularly Russia, Syria and Iran, who can help out. In testimony before the Senate Armed Services Committee in January 2017, the Director of National Intelligence and the director of the National Security Agency said we are not prepared to deal with war of this type. They certainly understated their point. This is a war we do not know how to win.

2

Information War

In 1995, Winn Schwartau, who had just written a book called *Information Warfare*, was walking ahead of us with our Canadian hosts. My wife had been listening to him speak part of the day, and she was anxious to draw some conclusions about what he had been saying. She said, kind of nonchalantly, "He doesn't seem to know very much about information warfare, does he?" Winn has sharp ears and stopped almost in mid-stride to wait for us to catch up. He ingratiated himself to my wife by starting with "young lady"; then he added something more unexpected. "You are right about that. I don't know as much about it as some of the people, like your husband, but I can do more in a day to wake people up than he can do in a year. I'm going to go to Europe tomorrow and we are going to hack some Defense Department computers, live, on national television." The guy knew how to make a point.

He wasn't waging war, but his demonstration did compel a few people to want to make changes to the way they did security for their systems in the Pentagon. It wasn't the massing of an army to attack the decisive points of a battlefield that can compel someone to act as you would want. He was using the Internet to communicate with masses of people about hacking in the information age. Some of those people worked in Defense and some of them on the Hill. This is not classic war, but it is Information War. You have to like the guy for practicing what he preaches.

Information war, sometimes inaccurately called cyberwar, is supposed to be a way of augmenting and enhancing other forms of conflict. It does not replace war, but it changes the way war is fought, as shown in these examples from a speech by George Tenet, then the director of Central Intelligence:

> For example, in an interview late last year, a senior Russian official commented that an attack against a national target such as transportation or electrical power distribution would ... by virtue of its catastrophic consequences, completely overlap with the use of [weapons] of mass destruction.
> An article in China's "People's Liberation Daily" stated that ... "an adversary wishing to destroy the United States only has to mess up the computer systems of its banks by

16

hi-tech means. This would disrupt and destroy the US economy. If we overlook this point and simply rely on the building of a costly standing army ... it is just as good as building a contemporary Maginot Line."

A defense publication from yet a third country stated that "Information Warfare will be the most vital component of future wars and disputes." The author predicted "bloodless" conflict since ... "information warfare alone may decide the outcome."[1]

The concept is remarkably similar to the one proposed by Major General A. H. Jomini, long before information war (IW) was ever contemplated. When IW was first defined, its use was limited because it was how the militaries of the world saw their role in fighting. Now some governments do not see the military as the only implementer of this kind of conflict and have expanded the scope and use of information war, even into domestic politics.

We also have a difficult time deciding who the enemy might be in these conflicts. We have trouble figuring out if they are at war with us, or with anyone else, for that matter. Terrorist groups, dictators, gangs of thugs and killers, thieves and extortionists can confuse us when we are at war, because we never know if we are at war with them or not. Sometimes they are just hanging out where a war might be; sometimes they are just trying to benefit from the confusion; sometimes they are fighting alongside one side or the other. Those of you who saw or read *Blackhawk Down* know what I'm talking about. It was easier when the armies lined up and everyone could tell that they were going to make war. "Shoot the ones in the blue uniforms," my Virginia ancestors used to say.

We are already at war, but all the confusion about what war has become has caused us to miss it. It isn't the information war that our military leaders had planned, and it doesn't look very much like a traditional war. Mr. Kissinger said that if we aren't careful, we will end up in Cold War with China, but he is probably too late with that kind of advice.

Information warfare started as a simple idea to use the denial, modification or manipulation of information to feed war operations. If we could blind a satellite, we could deny information about troop movements to our enemies. If we could change or deny military communications from leadership to soldiers in the field, it would benefit our combat operations. If we could sabotage radars, we could fly aircraft into another country without being seen. This view still envisioned mostly military against military conflict, with computers directly or indirectly engaged in warfare. There was also a view that combat was not required, nor necessarily involved, in executing some of the aspects.

The elements of information war are laid out in military doctrine long since forgotten by most military leaders. They define what is known as "total war," a concept accepted in military doctrine but seldom practiced the way it is defined. It includes economic warfare, the manipulation of information exchanged or used in trade, as an instrument of state policy; command and control warfare, which attacks the enemy's use of information to control and lead its military and support forces; psychological warfare, which affects the perceptions, intentions, and orientations of others; intelligence-based warfare, which is the integration of sensors, emitters, and processors into a system that integrates reconnaissance, surveillance, target acquisition, and battlefield damage assessments; electronic warfare, used to enhance, degrade, or intercept radio, radar, or cryptography of the enemy; cyberwarfare and hacker warfare. The latter two are separated in doctrine by only by a fine line, and I do not attempt to separate them here. I propose the addition of another type of warfare long known, but hardly ever talked about—political warfare—which was never in the domain of the military, thus not in the doctrine defined by them.

Information war has been defined differently over the years, and the definitions of concepts have changed with the expansion of information technology. In 1996, RAND called it strategic information warfare, a term that combines strategic warfare and information warfare. From this perspective a country "may use cyberspace to affect strategic military operations and inflict damage on national information infrastructures" where the damage may affect considerably more than military operations.[2] But the use of cyberspace as a vehicle for war goes further than this limited definition set out thirty years ago.

We usually think of war only as something done by the military, but that is a narrow view. In *The Changing Role of Information in Warfare*, Khalilzad and White tell us that information will make a military stronger by leveraging and synthesizing the capabilities that already exist.[3] That goes for terrorists too, of course. One of the Obama administration's big concerns is that the tactics we use in the information war will spill over into the world of terrorism. That kind of concern is too late, but better late than never, in this case.

Seven types of warfare are documented both by RAND and by Defense Department doctrine. Combining types of warfare is what some people call "total war"—the mixing of combat operations with disruption and control of economic and political functions of government. It is not only militaries that carry out this kind of things, but they are generally behind it. We are not very good at it, but the Chinese seem to have been paying attention, even if our militaries were not.

In 1995, two Chinese authors did a study of information war and defined it in terms used then by United States and Chinese militaries, to show how it was applied. What makes this report interesting is the description by Chinese military publications of what U.S. information war really is:

> Information warfare is combat operations in a high-tech battlefield environment in which both sides use information technology means, equipment, or systems in a rivalry over the power to obtain, control and use information.... We hold that information warfare has both narrow and broad meanings. Information warfare in the narrow sense refers to the U.S. military's so-called "battlefield information warfare," the crux of which is "command and control warfare." It is defined as the comprehensive use, with intelligence support, of military deception, operational secrecy, psychological warfare, electronic warfare, and substantive destruction to assault the enemy's whole information system including personnel; and to disrupt the enemy's information flow, in order to impact, weaken, and destroy the enemy's command and control capability, while keeping one's own command and control capability from being affected by similar enemy actions.[4]

What has happened since Baocun and Fei wrote their description is a gradual change in how information war is conducted. It has become a part of national strategies for political wars fought between countries, and between political rivals, to include those political battles at a national level. Information war is below the threshold of what governments, scholars, and the press call war, so it is ultimately used as a means to avoid war. It is changing and becoming a replacement for what we have known as war in the Uppsala University context.

In the 1970s version of information war, the sub-elements can be put in a more modern context.

Economic Warfare

Economic warfare is the manipulation of information exchanged in trade (either denial or exploitation) as an instrument of state policy. We usually don't think of this as war, even though it is. Wars are fought over who gets to trade with whom, and what they get to trade. Trade routes are big targets in a war, and we know that even a hint of trouble in the Gulf of Oman will raise gas prices in the U.S. by a dollar or more. But that is not the kind of thing this is about.

The Chinese use their intelligence services and military to collect information from the competition and feed that back into their companies. From a policy view, they steal information as a part of their national strategy to win an economic war. Their military still owns some companies and what they don't own, the Central Committee controls. They win bids; they control their own commodity prices; they harass the competition inside China. They steal intellectual property, which they then use to compete with the companies they steal it from. They leverage their surplus for political benefit and manipulate their currency valuation. They manipulate their laws to make a country comply with their objectives. They call this competition; we call it a few other things, but none of them are war.

Chinese companies are not what our Western world business experience would make of them. They look like our businesses. They have boards; some are privately held. They have articles of incorporation and bylaws, but they are still controlled by the government. Some of them are owned and operated by the People's Liberation Army, though that is less true today than it was five years ago. We see that difference in public very rarely, but it happened with Alibaba.

Anyone who follows the Internet knows about Alibaba and Yahoo! Alibaba is not some little Silicon Valley start-up. It has 23,000 employees in China, India, Japan, Korea, the U.K. and U.S., getting startup capital from Softbank, Goldman Sachs, and Fidelity. It includes Taobao Marketplace and Taobao Mall, which are the Amazon and eBay of China; Alibaba Cloud Computing, which does a cell phone operating system and is building the equivalent of Amazon's Cloud; China Yahoo!; and Alipay, the PayPal of China, which claims to be bigger than PayPal.

In 2011, investment companies were buzzing over Alipay, and for good reason. In the business world as we know it, corporate moves are decided by the board of directors, if a company is public, like Alibaba. Companies that own substantial amounts of stock generally have a say in these moves. There may be a board fight, but at least the board will know what is at stake and what has been proposed. Imagine Yahoo!'s surprise when Alibaba transferred ownership of Alipay to a Chinese company that is owned by the Alibaba Group CEO. They didn't find out about this for seven months.

This would be like Ford transferring the Lincoln division to another company, owned by its CEO, and not telling the stockholders they were going to do it. Alibaba claims it did this because of government regulations, but anywhere else in the world, the board would recognize the regulations and discuss what was going to be done with the company. The Chinese boards are not real boards of directors.

In 2017, the Chinese company Ant Financial purchased part of a company, Moneygram, that is famous for money transfers and loans for those seeking immediate cash. The Committee on Foreign Investment in the United States (CIFUS) considered the purchase, with regard to the fact that China ruled on Alipay without consulting anyone in the company. What Alibaba did was wake everyone up to the idea that the Chinese do not do business the way the rest of the world does.

Their government can, and will, act to control its business sectors whether the rest of world likes it or not, even saying that loans given to Chinese citizens must be made by business entities controlled by Chinese companies. Alipay could not be part of a business interest owned by a company in the United States. Indeed, in examining the proposed sale of Ant Financial in 2017 CFIUS took the same approach as the Chinese did with Alipay and did not approve the sale of Ant.

China has moved to making acquisitions of technology very early in the development cycle by buying start-ups as a rapid pace. Those acquisitions have been examined as part of a study directed by the Secretary of Defense of what technologies are being financed by Chinese businesses, which spent $9.9 billion in the U.S. in 2015. When the *New York Times* did a series of articles on this kind of investment, it found it difficult to find anyone in U.S. or Chinese businesses willing to talk about the amounts of money or types of technology involved in the investments.[5]

Command and Control Warfare

Command and control warfare (C2) is the attacking of the enemy's ability to issue commands and exchange them with field units; these attacks are called, in turn, anti-head and anti-neck operations. "Anti-head" and "anti-neck" are ridiculous terms that nobody uses in real life. All this attacking of people in command is not new. Every general who ever lived has known it was a good idea to kill off the other generals and officers who were leading the troops. Now, we can isolate them and cut them off, and it doesn't matter quite so much if they are dead. What is different about this is we might even need them alive, if we are issuing orders in their names. We want to isolate them and cut them off, but not kill them.

Early in the 2008 U.S. presidential race, the Chinese hacked accounts of the McCain and Obama campaigns, apparently looking for position papers and the directions the candidates would take. This is the kind of "looking around" that identifies the individuals who are writing the things the president will read and how they think. They hacked the accounts of congressmen and their staff members. Around the same time, they hacked the account of the Secretary of Defense. Two members of the House of Representatives, one of them my representative, said they were hacked by Chinese-based computers because of their investigations into human rights violations. Others, whose names do not appear in the press, have been hacked too.

The example of GhostNet makes this easier to understand. About all anyone can authoritatively say about this network was in two reports by the Information Warfare Monitor and Shadowserver Foundation, published a year apart.[6]

In the first report, researchers said they were not so sure that China itself was involved and that the spike in Internet hacking from China could be due to a 1,000 percent increase in Chinese users over the previous 8 years. In their April 2010 analysis, *Shadows in the Cloud*, the researchers had much more on how information was being stolen, what it was, and where it was going. They started looking at an example. The target was the Dalai Lama. The information being stolen was coming from Indian embassies in Belgium, Serbia, Germany, Italy, Kuwait, the United States, Zimbabwe, and the High Commissions of India in Cyprus and the U.S. Not very many ordinary hackers have an interest in the Dalai Lama.

The control servers for these attacks were in Chongqing, China; the Chinese are certainly interested in the Dalai Lama. The control servers used social networking sites, webmail providers, free service-hosting providers, and large companies on the Internet as operating locations and changed them frequently. They used similar, specifically targeted attacks against users, and collected 1,500 letters from the Dalai Lama's personal office. They also sucked out the contents of hundreds of e-mail accounts located in 31 different countries.

The difference in the accountability to China between the two reports was *attribution*, a topic beyond just cyber attacks. Once in a while false claims of credit take place in the cyber world, but most of the time cyber attacks are anonymous and denied by all involved—at least, that was true for the last 25–30 years. When Mandiant, Inc., produced a report called "Advanced Persistent threat (APT) 1," identifying hackers working for the Chinese army, and the Justice Department indicted five of them, there was official recognition of something most of the cyber world already knew: The Chinese government is behind many of these attacks that were targeting U.S. industries in almost every economic sector. That was powerful attribution.

At the national level, the U.S. Director of National Intelligence (DNI) laid out three things that are required to claim that one country hacked another.[7] He said we must know, first, the geographic location of the attack; second, the identity of the actual attackers; and third, the person responsible for actually directing the attack. An example of how difficult that is to apply follows:

In October 2012, Leon Panetta mentioned the possibility of a "cyber Pearl Harbor" to a group of U.S. businessmen in Washington, D.C., reopening a discussion of what an act of war really is in the cyber world.[8] So far, the only thing to have been publicly discussed as an act of war was an attack against the U.S. electric grid, so finding that someone might be preparing for that type of attack was disturbing. *The Times of Israel* reported the discovery of Iranian software in the U.S. grid, giving concern to almost everyone who knew about it.[9] There are several groups that question this discovery, the source of it, and whether it may have occurred. In the best of circumstances, there is always information that disputes the contentions required to make attribution. But, for this example, we can assume it did happen and we want to discover who did it, where he was, and who ordered it. We have incentive to do the work on this because it is a potential act of war.

Internally, governments want to decide whether the effort is worth the time and resources to discover what is needed to satisfy attribution. In order to constitute an act of war, the attack would have to have been likely to succeed in bringing down a substantial part of the electricity grid and harming people who would die or be injured by not having electricity. Hospitals, even those with emergency power, would lose some patients if that power could not be maintained. There is disagreement about whether or not this is force of a type that constitutes an act of war, so international lawyers get involved in this kind of decision. These types of debates are a little like discussions about how many angels can dance on the head of a pin, because we do know what an act of war is when we see it.

President Obama called the North Korean attack on Sony "cyber terrorism" rather than an act of war, legally appropriate given that it produced little physical harm to anyone. But North Korea was trying to impose its will on a business operating in the United States, something we should find reprehensible, criminal, and worthy of some retaliation. There is no end to this kind of activity unless there are consequences. It used a destructive attack that wiped servers belonging to Sony, but it was difficult to characterize that as a use of force in a traditional sense. Cyber terrorism is not thought of as war, so how is an attack on Sony that much different than an attack on the electric grid? There is no simple answer to that question because it is a policy decision and not just a matter of fact. Every government can make a different decision, sometimes even different decisions for similar events at different times.

Our ability to attribute cyber events to other countries is rapidly increasing, but we almost never respond in a way that convinces the ones perpetrating the actions to stop. We are getting better at attribution, but not better at retaliation or deterrence. That may be because we are not politically willing to do what has to be done to stop China, Iran, North Korea or Russia from continuing those attacks, but it may also be because the United States cannot do anything credible that will convince those countries to stop.

Hacking is a sophisticated business, but it is not a business the governments of countries are really in. What most governments do is intelligence collection or law enforcement. Their hackers are not trying to make money from the hacking they do, and they do not want to be discovered doing it. That is what makes the hacking by army units in China and the Russian meddling in the U.S. elections so odd. In both cases the United States has been able to attribute those attacks, with some certainty, to the two countries. In the Chinese case, the hackers were named and indictments brought against them, even though there was little chance of ever prosecuting them. That is good news—in a way—because attribution applies to more than just hacking.

If someone sinks a cruise ship in the Gulf of Mexico, we would like to know where the attack came from and who was behind it. Everybody will know it sank, but they might not know who did the deed. Before we attack Mexico over this sinking, it might be worth the effort to find out who did it. On the Internet, that can be harder than in the physical world, but not much harder.

Now, can we speculate about how those records from security clearance applications are going to be used? Yes, but it is only a guess. When anyone knows your past, they can influence your future. If they know things that you would not want the world to know, they can influence your future. They can plant information in those files that will show up during the next investigation. They can put whole records into the files, showing that someone has a security clearance when they do not. It is not what you think of as war, but it depends how that information is used. When a country steals your internal mail, it feels like they are after you. It can seem like war, whether scholars or political leaders call it one or not.

In seems almost intuitive that if there are nearly a billion Internet users, those numbers would make it impossible to find the people who are trying to get into information systems. In spite of the numbers, the intelligence services of several countries are doing it. Hackers are watched by other hackers, people in governments, and private-sector companies. So they have to take steps to make sure they don't get caught, or they cover their tracks so they can deny doing anything wrong. It takes time to find the right people, just as it took time to find out who was responsible for the 9/11 attacks. We didn't launch an attack on Saudi Arabia, even though most of the hijackers were from there. President Bush had to figure out who was really behind the attacks, and that took time.

The long process of identifying the source and method of the attacks on OPM brought the investigators to China. This is because investigators talk to each other, telling how one attack or another was done. When the Shadowserver Foundation found letters from the Dalai Lama that were stolen by China, investigators thought the Chinese were after a political opponent. But they found more than that. They started to see the same techniques, from the same locations, being used to attack other systems. There were 760 companies in all, and 20 percent of the Fortune 100. That is a scary number. This is the kind of attack, spread over several months, and extremely successful, that can get our leaders excited and ready to do something. It is right on the dividing line between war and not war, and that is where the Chinese like to stay.

Electronic Warfare

Electronic warfare is the practice of enhancing, degrading or intercepting radio, radar, or cryptography of the enemy. This is an old, very sophisticated, highly classified field that has had more success than most of the other areas of information war, yet almost nothing is known about how it is done, or where. That is a good thing. Hackers, today, use some of the same techniques, but few of them know that some of the things they have been doing have been done for a long time. Usually, if a government is doing it, it is classified national security information, and there will not be much to see of it. Generally speaking, if people are talking about it, it isn't the government doing the talking. Governments like to keep this quiet.

A good example is the Obama administration's consideration of using viruses to attack the radars of Libya's military.[10] *Wired* reported that the administration considered attacking Libya's radar sites but thought it might take too long to get the plan together and launch it. We don't see much like that in the press ever, and certainly not from any administration in recent memory. The obvious difference between what the Russians did in Georgia and what the *Wired* article was talking about is the military-on-military aspect of it. No country likes to talk about this kind of thing, and no administration ever should.

Starting in November of 2010, several systems were hacked by someone who established over 300 control systems, almost all around Beijing. What made this different was that the attackers were going after a place called RSA that was famous for its ability to do encryption of various sorts. RSA makes a token that many in business have seen. A user logs onto a home network and the software asks the user to type in a long number that is read from the token. It is just that one time, and it changes, so it is not the same number the next time. The nice thing about the RSA token method is you can be pretty sure anyone who logs in with it is an authorized user. We would probably think a place that makes security devices would be secure, but over the past couple of years, more than one of them has been successfully attacked. The people doing it are good.

During the next few months, several other major companies were hacked, and there was a pattern to these that will make anyone nervous who sees the list.[11] There is the IRS, Verisign (another crypto-solutions company), USAA, which primarily handles insurance and banking for military people, several locations of Comcast and Computer Sciences Corporation, a few locations of IBM, the U.S. Cert, which handles investigations into computer incidents at the federal level, the Defense Department Network Information Center, Facebook, Fannie May, Freddie Mac (just so we have most of those housing loans covered), Kaiser Foundation Health Care System, McAfee, Inc. (the virus people who do nearly all of defense networks), Motorola, Wells Fargo Bank (and Wachovia, now owned by Wells Fargo), MIT, University of Nebraska–Lincoln, University of Pittsburgh, VMWare, the World Bank, and almost every telecommunications company of any size, anywhere in the world. That last one included all the major telecoms in China. So they are hacking their own telecoms. It is almost like someone said, "Go out and get everything you can." There are probably some that have yet to be discovered.

Intelligence-Based Warfare

Intelligence-based warfare is the integration of sensors, emitters, and processors into a system that integrates reconnaissance, surveillance, target acquisition, and battlefield damage assessment. These techniques can be used to both seek and hide assets. Iran and North Korea have taken to building their sensitive sites underground, which certainly is not new, but is a recognition of how effective other countries have been at discovering their capabilities. If a country wants to hide what they are doing, it costs them quite a bit in resources to do that.

Psychological Warfare

Psychological warfare is the use of information to affect the perceptions, intentions, and orientations of others. Psychological warfare is probably the oldest and best-known form of information war, and it is at the root of political warfare. Think Tokyo Rose and her radio broadcasts to GIs in World War II. When the Chinese lowered the value of their currency after warning us about putting more restrictions on trade, they were making a point that they had the ability to disrupt our economy. They don't have to do anything else for the psychological effect to take hold.

I was surprised to see a picture of a J-20 stealth fighter in the *Wall Street Journal*. The *Wall Street Journal* is not known for its aviation reporting, nor for digging up important news on stealth fighters in all the different countries of the Asian Pacific. It turns out that the U.S. Secretary of Defense just happened to be visiting the Chinese that day, and this fighter was sitting out on a runway where anyone could see it. It taxied around a little while to make sure nobody missed it, then took off and flew around. Lots of pictures were taken, by everyone present, and sent to newspapers and magazines over the earth. How odd.

Since I spent most of my military life protecting weapon systems so they couldn't be seen before they were operational, I know the principle involved here. It takes a lot of money and time to get a stealth fighter to fly. Before anyone in the world sees it, no military force on earth wants another military to know that it exists. They hide it in hangars, fly it at night, or do other things that make it more difficult for people who look for stuff like that to find it. They don't want people to see it until it gets to the point where it can be used for something—usually an operational mission where it collects intelligence or bombs something. When other militaries see it, they are going to want to start working on something better, or on some way to counter its capabilities. For a few months, maybe a year or two, the advantage is useful to those who have it, but it will eventually be overtaken. It is a kind of game, but a serious one.

We saw the same thing with the Chinese aircraft carrier, the first they had ever set on the water.[12] The BBC called this the "worst kept secret in naval aviation history." The Russians built it, Ukraine sold it to China, and China said it was going to be used as a floating casino. When they got it into port they started to work on it, and it was obvious it was not being equipped as a new casino unless the customers could land on a moving deck. You can't hide an aircraft carrier while it is sailing around on the ocean, but you can hide it while it is being built. They haven't been trying to hide it or even conceal some of its most important capabilities. They could have said they were building a test vehicle or a cruise ship, but they didn't even want to do that. This is openness that is curious. The second aircraft carrier was launched in March 2017 and shows the influence of design upgrades and innovation lacking on the first. There are three more on the way. The Chinese learn fast, and they are not spending extra money trying to hide what they are building.

These are images of what might be a step towards war. It isn't war; it is just a picture of what could happen in war, and it was intended for a number of different audiences—their own and others. The Chinese stealth aircraft was one of a kind, but it served its purpose. They wanted us to guess what they could and would do. *Jane's* published a picture of a cave entrance where Chinese submarines were going into and out of the base of a mountain on the coast. There could have been 100 of them in there, or just a couple with the numbers being repainted each time they come and go. It was deception—all part of war. The Internet will allow them to get these images out to large numbers of people and not just to the intelligence services of people who are spying on them.

They could keep quiet about them, and only the intelligence services of the world would know much about what they can do—only they didn't. There was no filtering by the intelligence communities of other countries. The Chinese didn't try to hide these things and release images of them at times when there were things going on in the world that could be affected by them. The first case was during the visit of the Secretary of Defense; the second was during the testing of a new president; and the third was a few months after the ASEAN Forum, when the U.S. was invited back into the region. At least for now, the Chinese find it more useful to make images of war than to make war. They have been doing this for some time.

Cyberwarfare

Cyberwarfare is the use of information systems against the virtual personas of individuals or groups. This is attacking people, or groups, as they exist in digital form. It is a little bit like the movie *Avatar*, except machines can pretend to be one person, or multiple people at the same time. In a virtual world, a machine can be a person and can function the same way. It is possible to be more than one person at the same time. It can be difficult to find the real you, and that may be what is intended. It is easy for someone to pretend to be you and act as you might. We seem to get e-mail, advertising Viagra, from some of our best friends. This is somewhat the idea of information war. Cyberwarfare is usually used in conjunction with one of the other types of information warfare, and is the newest form of it, but it should not be confused with hacking for a criminal purpose. It is more sinister than that. Besides the potential to undermine basic military services, people are hacking our banking structure, home loans, electricity grid and lots of other things. But they are also going one step further. The Chinese are trying to undermine our telecommunications infrastructure by buying into it where they can, and having a substantial market position in equipment where they cannot.

Most of the world's telecommunications infrastructure is owned by about 50 companies. AT&T is still the largest, by revenue, followed by Vodaphone in the U.K., Telefónica in Spain, China Mobile, Nippon Telegraph and Telephone, and Verizon. These are all global carriers, so they operate in quite a few countries and have agreements with the other carriers to swap services where they need access. They overlap in some countries but not all. Every country, for national security reasons, has some limitations on what ownership another country can have in its infrastructure. China enforces theirs, as do we. We all do this in the name of national security.

The Chinese are trying to buy into the U.S. infrastructure using, among other companies, Huawei, and the U.S. has not been willing to let them do it. In the past few years, Congress and the Committee on Foreign Investment have intervened in Huawei's attempted purchase of 3-Com, 3-Leaf, a piece of Motorola's network infrastructure (later sold to Nokia Seimens) and Emcore, a New Mexico–based company that sold fiber optic equipment. It is clear our government does not intend for them to be successful, and Huawei has finally realized it. They have finally stopped trying, and other companies have taken their place.

Both Huawei and ZTE, the second largest equipment maker in the world, were excluded from a Sprint/Nextel bid, and several U.S. senators were said to have sent a letter encouraging them to be denied that opportunity. They were stopped in similar bids on AT&T networks and 2Wire, a U.S.-based company owned by Pace, a British firm. 2Wires's main business was residential broadband. The Commerce Department recently said that Huawei was not going to be allowed to bid on a contract for a national wireless network for first responders, because they suspected Huawei was linked to the Intelligence Services of China.[13]

Another indicator of how that business connection lies in the use of ZTE to violate sanctions that China voted for in the United Nations. The violations were exposed in the interesting export violation case of ZTE. Sanctions were levied on ZTE that would limit its ability to buy technology in the U.S.[14] The Commerce Department, which enforces these sanctions, said ZTE acted "contrary to the national security and foreign policy interests of the United States. ... Authorities allege ZTE broke export rules by supplying Iran with U.S.-made high-tech goods and said they uncovered plans by ZTE to use a series of shell companies to illicitly re-export controlled items to Iran in violation of U.S. export control laws."

The Commerce Department published internal documents of ZTE Corp marked "Top Secret, Highly Confidential" to substantiate its claim that ZTE knew what it was doing when it funneled hardware and software from Microsoft, Oracle, IBM, and Dell to Iran. The first document, titled "Report Regarding Comprehensive Reorganization and the Standardization of the Company Export Control Related Matters," clearly shows that ZTE knew what the export rules required, and knew there would be trouble if they were discovered trying to skirt them. "Group Z" described in this document refers to North Korea, Vietnam and Cuba. ZTE, at the time of writing, was exporting U.S.-produced products to Iran, Sudan, North Korea, Syria and Cuba. They outlined the methods used to avoid detection in all of these countries, summarized in the second document, "Proposal for Import and Export Control Risk Avoidance."[15] The U.S. sanctions had been in place for less than a week when they were withdrawn by the Obama administration, without explanation.

We got a view into the quick withdrawal of sanctions against ZTE and the way ZTE sought to resolve the issue with the U.S. government. ZTE started by announcing what it would do in order to mitigate the actions taken by members of its board of directors to sell embargoed items to Iran.[16] In the U.S. these kinds of actions would be severe and impact the board in ways that would resonate for a long time. We tend to see China's actions in the same light, when the government systems and the consequences are not the same.

Hackerwarfare

Hackerwarfare is the use of techniques such as modifying software to destroy, degrade, exploit, or compromise information systems, both military and civilian.

Information war is only part of war, but it is a part we can relate to. We rely on information systems to do most of our work and social contact, and it looks like the Chinese see that as important to their ability to fight. The Chinese steal two things that are related: source code for software, and code-signing certificates. These two together allow them to set up networks that look legitimate but are controlled by them. Original source code is proprietary to a company, but the Chinese demand it be produced under their "cybersecurity" laws. The software must be "secure and controllable," and to prove it, some companies are being required to submit source code. Some, have resisted, though the struggle is getting harder today than it was five years ago.

One vendor explained to me that the company resisted turning it over and resisted putting hardware or software in spaces controlled exclusively by Chinese officials.

Judging from what businesses have told me, they know that if they put their equipment in certain areas, it will be stolen. So, if the Chinese cannot get code from the company directly, they steal it. That has been going on for some time. China announced that enforcement of the cybersecurity policy was to begin in June of 2017, but had actually been applying it long before.[17] Imagine if any other government decided to have every foreign business like SAP or Airbus submit code to a special service for review, then turned over that software to the competitors of those businesses. There are not enough lawyers to cover all the lawsuits that would result. But China seems to be able to get away with it because they allow companies operating in China under Chinese law access to some of China's billion potential customers.

Having that source code and the security certificates that goes with it allows the Chinese to produce software that will collect intelligence information of all kinds. It looks valid, and users will never know the difference. On a large scale, that could be interesting. Parts of the world's networks are controlled by China; parts of others use components made by Chinese electronic companies; China has demonstrated a capability to manipulate networks on a large scale.

Several years ago, for 18 minutes, quite a bit of network traffic was rerouted to China.[18] Most of it was from our Defense Department. Was it an accident? This tickles my imagination because it doesn't seem like something accidental—although since it happens in various parts of the world, on a regular basis, it may be. It is also possible that it was just practice for something bigger. China can, and does, manipulate network components to satisfy military objectives.

Increasing Capabilities for War

At the same time that the Chinese develop their strategies for information war, they are strengthening their ability to fight real wars by increasing support for space operations, nuclear combat capabilities, and cyberwar.

China had 15 space launches in 2010, a national record, and they have a program to get something on the moon. That was the first year that they equaled the U.S. in launches. They have developed anti-satellite weapons and may have intentions of using them against both communications and spy satellites. They practiced by using one on an old Chinese weather satellite, and the U.S. has accused them of using lasers to blind our satellites. They have just launched their fifth GPS satellite, which can mean a number of things, but mostly that they want to have their own, rather than using someone else's. They may want to have some, if they decide to shoot all the others out of the sky.

They have stepped up their espionage. In the past 15 years, China has stolen classified details of every major nuclear and neutron bomb the U.S. had in its inventory.[19] They have had ongoing espionage activity at the nuclear laboratories—Los Alamos, Lawrence Livermore, Oak Ridge, and Sandia— that produce and develop the weapons. This allows them to make their weapons smaller and easier to shoot a long way on a missile.[20]

China has stolen U.S. missile guidance technology and exported it to other countries such as Iran, Pakistan, Syria, Libya and North Korea. They sold medium range missiles to Saudi Arabia and they trade extensively with Iran, which is not our best friend after trying to get Mexican drug gangs to hit embassies in Washington, D.C.[21]

There have been accusations that the Russians and Chinese have planted software in our electrical grid that will give them control of it if they want to use it. They have been accused of introducing counterfeit servers and other Internet equipment that appears to belong to legitimate companies but was not made by them. They are probably doing all of these things and quite bit more. Nobody is going to produce proof of it, since it could be right on the verge of war, and it is the kind of war the Chinese could wage, but we could not.

Besides chips, the Chinese make quite a few other network components, but chips seem to be an interest that exceeds the others. In 2015–2016, the Chinese bought 12 U.S. chip-making companies and tried to buy 6 others. Those that were unsuccessful were rejected for business reasons or came under scrutiny from the Committee on Foreign Investment in the United States (CFIUS).[22] The Chinese now use various companies to make the purchases so it is not as easy to discover the state connection. Controlling the chip markets puts the core technology for computers in the hands of the Chinese government. But that is not the only aspect of networks that the Chinese control.

Several types of hard disks are made in China, including HP, IBM, and Western Digital, some of the most used storage drives in the world. Chinese companies make workstations, PCs, wireless equipment, routers, servers, DVD players and recorders, motherboards, cables, the test equipment we use to test all of these, and 300 antivirus products. China has business ties (partnerships, teaming arrangements, or joint ventures) with most of the major U.S. antivirus companies. One source told me that the Chinese were eager to help her prominent company write the source code for new virus software, which her company allowed. So, that U.S. company had software produced in China running on computers in the U.S. Their customers would have no idea that China was involved in making their antivirus software work. The security team was uncomfortable with that, but senior management told them they were overreacting.

If you want to build a network of computers, the Chinese can help you get all the things needed to do that, and a company that will build it for you. Even if you get another company, it would be hard to avoid buying components that were not made in China. This gives them leverage and potential for controlling large portions of the Internet in war. Imagine the Stuxnet worm in big numbers, everywhere. The Chinese have that capacity now. They have the ability to shut down the Internet in time of war.

What the Chinese have the potential to do is increase Stuxnet by a factor of 1,000 or more, and they have the delivery mechanisms already in place. They already make them and can put worms or special codes in almost anything, then wait for the right time to turn it on. Stuxnet turned out to not be as controllable as it should have been, but in an all-out war, this will not matter very much. As an opening round in any other type of war, it might work pretty well. It can stop communications, or slow them down enough to open up other avenues of attack. The U.S. does not just sit around thinking about this.

In 2014, David Sanger and Nicole Perlroth from the *New York Times* published a lengthy story on the underlying issues related to the U.S. concerns over Huawei.[23] According to the story, the U.S. had concerns about the Chinese ability to insert intelligence collection capabilities into equipment manufactured in China. Huawei had already been stopped by CFIUS from procuring companies and technology in the U.S. because of its suspected associations with Chinese intelligence agencies. Huawei denied any associations with them.

According to Sanger, the National Security Agency (NSA) hacked the internal networks of Huawei in China and monitored the communications of Huawei's senior officials. It also exploited some of the servers that Huawei was building so that they could get access to China and to Huawei's customers through their world-wide equipment sales. After that, there would not be much reason to guess about the associations between Huawei and Chinese intelligence; the NSA would know the extent of them. Had it not been for Edward Snowden, this operation might never have been known. But Huawei still denies any association with intelligence services; the NSA does not comment on any operations it undertakes; and the users of Huawei's equipment are left to speculate about which governments are monitoring their equipment. Notice that no accusations were laid on any country's door for either side of these activities because they would both be seen as "in bounds" for intelligence collection.

Many years after the RAND paper on information war, the Joint Chiefs of Staff (see Figure 1) defined "information operations" in such a way as to show the linkage between the electronic mechanisms of computer networks, indicating that the targets of these operations are key influencers, mass audiences, and "vulnerable populations," a term not defined.[24] This is far removed from the original definition, which focused upon the use of computers against an enemy's computers and information sources. As the doctrine evolved, it tended to become an approach using the *persuasion* of human beings rather than the *force* typically used in combat.

In 1993, an incident with a warlord in Somalia killed 18 U.S. Special Forces soldiers and generated a study group called the Aspen-Brown Commission the year after.[25] The commission set out this definition of information war: "activities undertaken by government, groups, or individuals to gain electronic access to information systems in other countries … as well as activities undertaken to protect against it."[26] The commission laid out a role for intelligence services, especially the Central Intelligence Agency, in information war. That role was largely defining and examining the intentions and capabilities of other countries.

But there is a little more to that role than just observing and reporting. The CIA and the National Security Agency (NSA) conduct covert operations in the furtherance of U.S. interests.[27]

In Russia the Security Services, the FSB and GRU engage in this kind of activity. In China, it is the army and the intelligence activities. These are the agencies, augmented by military forces, that are most involved in information war. Covert portions of war are run under the authority of these groups and often with their direction. Covert programs have to have plausible deniability to be successful; i.e., a government must be able to say that it was not involved in a particular activity. The denials are frequently not very credible, in spite of considerable efforts to make them appear so.

While countries may all do similar things, one country's actions do not necessarily appear the same to the others. Attempts at influencing dissidents in one country can be seen as fomenting insurrection, something most of us can understand. We tacitly accept these kinds of persuasive actions because most of them can be ignored by an informed public, yet some governments use social media, media, and monitoring to determine what limits are placed on being informed. They encourage informed individuals to limit their sources,

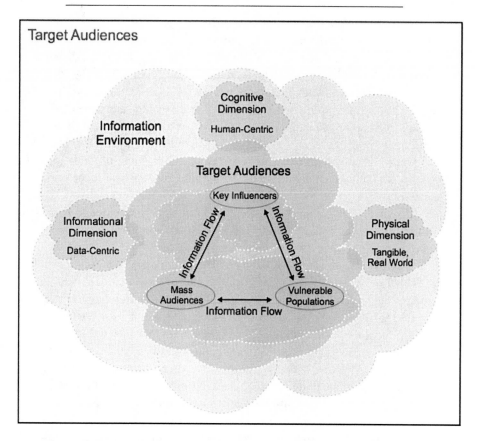

Figure 1. Information operations emphasis on persuasion of noncombatants and key influencers as part of warfare (U.S. Joint Chiefs of Staff).

and not all of those individuals are inside their own countries. Governments push those limits to extremes to influence perceptions of people to those favored in narratives created by leadership. But that is not all they do.

In preparation for a possible war, governments have to do many things that, if discovered, can be interpreted as either extreme provocations or acts of war. The act of planting software in an electrical grid that will disrupt service when commanded to do so falls into this category. While the software may lie dormant for many years, the fact of its being inserted can be seen as an act of war. Triggering that software is certainly war.

Although these kinds of activities are separate from intelligence collection and analysis, the grey areas between government covert operations and intelligence gathering are large in cyber and hacker warfare, where the skill sets are similar.

Reconnaissance done in preparation for cyber operations requires the attacker to do a certain amount of exploration of potential avenues of attack. We know that the Chinese hackers who stole the security clearance data from the United States had been in the system for three years, and the Russian hackers in Ukraine did not cause an outage of electricity without some preparation. The actions had to be tested, refined and executed at a time when the impact would give a benefit to those doing it. While they were in these systems, they could add records, delete them, or modify some aspects that could be "found" later by legitimate users including those doing reevaluations of persons seeking security clearances. In Ukraine they could just turn off the electricity.

If such code is discovered by the target before it is executed, it could be removed, searches could be performed for other copies and variations, and analysis could be done to determine the origin. Seldom is it a good idea to provide this information to third parties who might make the discovery public. Third-party analysis of code found "in the wild" can spread techniques used by the developers into every corner of the Dark Web, where it is refined and sold to others, often with improvements that make it more effective. But the U.S. and the Ukrainians had other choices about how they would respond.

Anyone discovering the software could modify it so it would not be effective, leave it there and never say a word about detecting it, or make it public to warn others that the attack is occurring and to notify the developer that there might be retaliation to follow. That choice is also a political decision.

Disruption of the electrical grid might be an act of war, though to Ukraine there are other acts of war that make this one less significant. In addition, there is very little Ukraine can do about it, even if the country does interpret it as an act of war, since it is a small country with limited cyber capabilities. It would be reluctant to take on Russia, assuming Ukraine could determine that the Russian government was behind the attack. That makes alternatives that do not disclose the event more attractive. But if the U.S., France, or Germany made that same type of discovery, the country that put it into those computer systems might find those countries can and might retaliate.

What deters the Chinese and others is the United States' capability, through its Computer Network Operations, to infiltrate and command the networks of other countries. Though there is no evidence it has ever done so, with the possible exception prompted by White House mention of the North Korean retaliation, having the capability is a strong indicator to adversaries that there are limits to how far intrusions into U.S. networks will go before retaliation may take place: "The Joint Information Operations Warfare Command (JIOWC) and the Joint Functional Component Command for Network Warfare (JFCCNW) are responsible for the evolving mission of Computer Network Attack."[28]

According to John Lasker,

> The exact capabilities of the JIOWC and JFCCNW are highly classified [state secrets], and DOD officials have reportedly never admitted to launching a cyber attack against an enemy. However, many computer security officials believe the organization can destroy networks and penetrate enemy computers to steal or manipulate data, and take down enemy command-and-control systems. They also believe that the organization consists of personnel from the CIA, National Security Agency, FBI, the four military branches, and civilians and military representatives from allied nations.[29]

But more complications are involved than just direct retaliation for a single act. At the same time the attacks on the electricity grid were taking place, the Russians were spying on politicians, government agencies, and private businesses in Germany, NATO countries, French TV5 (taking it offline for a brief period), Ukrainian government leaders, Russian dissidents, and the Dutch Safety Board, which was writing the analysis of the downing of the civilian airliner over Ukraine.[30]

These overlapping initiatives expose the Russians to detection by a number of security companies and other governments looking for evidence of hacking of any kind. The same thing happened to China when they overexposed their theft of proprietary information in the period from 2010 to 2016. With that many people looking for the groups behind the incidents, intelligence services find it more difficult to do their primary job of intelligence collection. They spend too much time denying attacks, retooling software, and covering their tracks. Apparently, however, this does not make them stop collecting. For the most part, offensive cyber operations are not being detected often enough to discourage the behavior.

Cyber intrusions are both an effective tool and one difficult to retaliate against. Although the Obama administration took the view that other forms of retaliation would be equally effective at deterring attacks, that decision is a political one. There has to be a range of potential ways drawn up in advance and prepared for. The potential targets would be as numerous as the potential targets for attacks on our country. Countries have attacked financial systems, businesses, telecommunications systems, hospitals, schools, government offices, public facilities, and utilities. It is extremely difficult to prepare for attacks on all of these kinds of facilities, so we have to choose which ones we want to concentrate on. That limits the "response-in-kind" options and forces a country to make choices about the other available options.

As the Director of National Intelligence said in 2016, the decision for retaliation is a political one.[31] Deterrence is the ideal situation, though some countries will not be deterred by anything other countries do. China is careful to use North Korea as its stalking horse in attacks on Sony and South Korea to avoid being the target of retaliation. North Korea does not seem to fear any kind of retaliation, so it is not deterred. In the same way, Russia uses hacker groups not directly affiliated with the state. Both of these provide plausible deniability for some types of attacks coming from the state. But this kind of warfare is far more complex than just cyber attacks on one another.

3

What Has Become
of War?

Sun Tzu, one of the most-read Chinese military leaders, was a proponent of winning without necessarily fighting. He said, "Hence to fight and conquer in all your battles is not supreme excellence; supreme excellence consists in breaking the enemy's resistance without fighting."[1] This is the ultimate war that those soldiers in Antietam and Verdun would have appreciated.

China has claimed the South China Sea, parts of which are also claimed by other countries. There is no fighting preceding it; no battles between armies; no real exchanges of missiles or bombs—at least, not yet. Governments involved in these territorial disputes do not say they are at war; they are just disputing territorial claims. A country takes over territory against the will of those with claims to it, yet it is hard to find a word describing what has happened in those places.

By international standards, the U.S. and China should already be at war. The U.S. has moved warships into maritime territory claimed by China. If China really believed it owned this entire expanse of land and water in the South China Sea, it could not believe the constant ship movements were innocent passage. "Innocent passage" is defined in the United Nations Convention on the Laws of the Sea, which lays out (in extreme detail) when ships of one sovereign can pass through the waters of another. Even by the terms of the convention, those ships are far from innocent. They challenge China on its claim to the sections of sea that they are trying to make part of China. That includes the island of Taiwan and might also include South Korea. China should interpret ships entering into that space as an act of war.

Seeing those ships, China could give consideration to declaring war or responding in kind, but no such thing has happened. It has responded with Russian-Chinese joint exercises and a few coast guard ships in disputed areas. In an area southwest of Hainan is the Tonkin Gulf, where the United States had disputes with North Vietnamese ships before the Vietnam War got under-

Figure 4. U.S.S. *Carl Vinson* enters the South China Sea (U.S. Defense Department).

way; there, the Chinese were running live fire exercises and have told other countries not to come in. This is a difficult order to defy, since all that is required to run live fire exercises is ships, ammunition, and communications of the intent. The ships are there all the time and the announcements could become more frequent if need be. Slowly but surely, the Chinese are demonstrating dominion over that water, and they are doing it in small increments that are adding up. These events are interpreted by government spokespersons who say these are not acts of war, just normal events that occur every day in world affairs. Much like the invasion of Czechoslovakia by Germany, there will be a time when that explanation will not be enough, but it will be too late for peace then.

At the same time, China wants those who listen to them to believe war *could* happen, and makes official statements to support that belief. The audiences for these kinds of statements are other governments, not the general public. The U.S. ships in the South China Sea were said to provoke a statement by some military officers that the U.S. should "get a bloody nose" for bringing those ships into Chinese territory.[2] Stories prompted by China's state-managed press said people in China smashed iPhones and picketed U.S. business outlets in protest. If we think about this, nobody actually does these kinds of things in reaction to a far off incident of mostly minor consequence to individuals. If we went out on the street and asked the average person in the U.S. what kind of action they would like to see their government take because of claims made in the South China Sea, most of them would ask for a repeat of the question. None of them would want to smash a $730 (in China) iPhone in protest unless it was gifted to them and cameras were rolling. However principled they might be, unless there was some political or personal benefit, they just are not going to do it. Governments are different.

Governments do what they can to promote their own views. While it might include handing out iPhones to smash, the Chinese government indicated that doing so was counterproductive.[3] In other words, they do not want the behavior to continue or spread beyond what the government can control. Both sides called on an old standby and have moved towards joint exercises with other military forces. The U.S. has done exercises with the Philippines, South Korea and Japan. China has done them with the Russians and others. This is the usual routine of gathering sailors, airmen and soldiers together and running around in territories claimed by both sides. They fire off live ammunition to prove they have it, but they would not fire it at other forces. This is a demonstration of capability. Very few countries take the exercises seriously because they are not indicators of friendship between the owners of ships who exercise together. The activity is not fighting, either.

China does not like to respond to aggression unless it has an advantage, but that does not have to be a tactical advantage in an engagement. In May of 2016, Chinese fighters harassed a U.S. spy plane flying near Hainan, which just happens to sit across from Vietnam. This was a few days before the president of the United States flew to Vietnam and announced the end to an arms embargo that had been a fixture of U.S. policy. President Obama says ending the embargo has nothing to do with China, but it is clear to everyone that China and Vietnam disagree about who owns certain islands in the South China Sea. Vietnam has decided to duplicate the Chinese actions and start dredging an island of its own in the South China Sea.[4] The Chinese set up missile launchers in May 2017, but these are not the anti-aircraft or ship missiles they have set up in the past. These are directed against divers in the water off the islands.[5] It is difficult for us to believe that militaries of the world even have such a weapon, but the Chinese do and Vietnam knows it.

The last time the U.S. and China clashed over these territories, the Chinese sent ships through the Bering Sea into maritime territory of the U.S. The president just happened to be visiting Alaska the day that happened. We should not believe that either the incident at Hainan or the one in Alaska was a coincidence.

When the U.S. deployed ships to the South China Sea, and China deployed ships to Alaskan territory, they both knew what they were doing. They knew the other side could interpret these as acts as warlike, but would not. The Chinese claimed innocent passage, sailing through the 12-mile limit, and the U.S. accepted that claim, even though the Chinese ships would have been an indirect threat to *Air Force One*, the aircraft used by the president of the United States. The U.S. sends ships into the South China Sea with the same kind of reaction. We can observe that neither side speaks of this as war, even though, by some interpretations, acts of war might have been committed by both sides. These tit-for-tat exchanges are only beginning.

The Chinese do not accept the standing of positions that are different from their own, and they believe the South China Sea is theirs. China did not participate in the United Nations discussions about the South China Sea. The U.N. Arbitral Tribunal ruled against them in July 2016, but China continues to act as if it has already won this battle on its own terms. It chooses to negotiate with the Philippines behind the scenes with no other parties present. It is building islands out of nothing to set an interesting precedent, an issue in the case brought to the United Nations. The second finding of the arbitration tribunal was that the exclusive economic zone (EEZ) of the Philippines was being interfered with by China, potentially a more important issue than the broader claim to the South China Sea. The tribunal said China violated the Philippine's exclusive economic zone by interfering with fishing and petroleum exploration, and by constructing artificial islands. The tribunal is saying that the building of those islands was part of the claim at issue. China will not negotiate with other parties that have competing claims, unless discussions are on their terms and yield their results. They continue to deny that the United Nations has any authority to make a ruling on this matter; rather, they negotiate separately with the Philippines' new president.

What the Chinese did after the ruling says more about them than what happened before it came. They ratcheted up the rhetoric, doing what they are good at: creating an image. They want that image to reflect what could happen if the "negotiations" with the Philippines do not go well. But if the Philippines won the day with the decision, why were they negotiating anything? The Chinese ambassador to the Philippines told the newly elected president and his minister what to say and not say about the ruling by the arbitration tribunal.[6] China's action was not misinterpreted by the Philippine president, but it was obvious he did not want to share what was said, nor his reaction to it, with others. He will begin more formal negotiations with China in 2017, and in the meantime, he is collecting Chinese money for various causes including the rebuilding of his home town.

China started to call the flights of military aircraft—which had been going on for over a year—"combat patrols." They built military-style hangers for the aircraft. They put military radars on oil platforms. They announced that a new satellite would be used to monitor "sea interests," when it probably will be used for a number of things besides that.[7] They put surface-to-air missiles on some of these islands, then took them away shortly after they were sure they had been seen by anyone who cared.[8] They put J-11 fighters on the same island. China's controlled press started to publicize stories of the iPhone smashing and the picketing of U.S.-owned establishments. They were not doing anything they had not been doing before the announcement of the finding by the tribunal, but they worked hard to have the world see it as an escalation caused by a conflict—never mind that it was one China had created. Truth in information war is always smothered by repetition and buried in counter claims and denial.

Denial is a major political component of this new form of war. If we were to believe what the U.S. and China diplomatic corps say, the countries of the world are peaceful, friendly and agreeable; they want the world to understand the issues will be resolved peaceably. China is helping the U.S. with North Korea. We would be surprised if those Chinese and U.S. aircraft and ships engage each other in the South China Sea. Were that to happen, we would find it more difficult to remember we are not at war. What has allowed us to do that in the past is a kind of logic called fragmentation by decomposition. We break down the issues into small parts that can separately be excused as events that are not war, where taking those same events together we might come to a different conclusion. Politics drives the deception, and it happens more than we realize.

During my six years in Ballistic Missile Defense, the Terminal High Altitude Air Defense (THAAD) was a system we hoped would shoot down ballistic missiles that were coming from North Korea, destined for the U.S. If that 20-year-old dream sounds familiar, it is still in discussion today. In briefings to Congress our leaders often said the main concern was North Korea's ability to mate a nuclear warhead to a long-range missile, at that time the Taepodong II. The website of the Federation of American Scientists lists the range of this missile at around 2,600 miles, not long enough to reach the continental United States.[9]

At that time, politics played into any comment being made about the North Korean capability. The Clinton administration was concerned, but did not want to say, that North Korea could hit the United States with a nuclear missile. Nobody said North Korea did not have nuclear weapons; it was known that the regime was working on them.

In fact, North Koreas missiles could hit Alaska and Hawaii, both part of the U.S. Twenty years ago, the director of Ballistic Missile Defense Organization stated the truth that the missile could not hit the continental United

States, a comment that was then misquoted several times by congressional leaders as "could not hit the United States." Senator Daniel Inouye spoke to one of his colleagues at a hearing discussing the issue and said, "May I remind the gentleman that Hawaii is in the United States." Nobody laughed. It was not the narrative the Clinton administration wanted to hear.

The debate was always about money then, how much to allocate to ballistic missile defense, particularly National Missile Defense, which was eventually built on a smaller budget because of the way the North Korean capability was described. That capability has come to be considerably greater than what was projected.

Now, nearly 20 years later, we pretend that situation is still the same. We want to install THAAD in South Korea, and we have actually shipped the missile launchers and announced they were operational. The Chinese claim to look at that deployment as a chance for the U.S. to threaten the effectiveness of Chinese missiles, an unlikely scenario. China is not the one threatening to destroy cities in the United States with a nuclear missile. The Chinese say the U.S. should negotiate (actually saying the U.S. should "do its part") with the North Koreans directly, and in exchange, China would cut back on the imports of coal, exercising some leverage over the economy. The Chinese bulk up their purchases of coal to the U.N. quota then declare they will comply with those quotas. They have not stopped North Korea's nuclear weapons program nor its refinement of missile systems to deliver them. The Chinese still use North Korea as a proxy for provocative actions that keep parts of the world on a defensive footing.

There may be some truth to what the Chinese say about THAAD, but as in most of their arguments they are very thin on facts to support their rhetoric. THAAD would work just as well on a Chinese missile as on a North Korean missile—probably better. THAAD interceptors operate at a high altitude, and the higher the altitude the better for targeting. Since the North Koreans want to target the U.S. and would not like their missiles to be intercepted, they are testing submarine-launched missiles, which are hard to hit with anti-missile systems. They come out of the water closer to their target and may not fly as high into the atmosphere. There is less time to make a target identification and launch a missile at their missile. It is hard enough to hit a missile with another missile without reducing the amount of time it takes to do it. The Chinese, North Koreans, and the U.S. know that. The North Koreans launch four or five missiles at one time to demonstrate they could overwhelm THAAD with numbers, but it is less than clear whether the North has just used a good portion of its inventory of boosters to show what it might do.

Travis Wheeler, writing for the *Diplomat*, said the Chinese have only about 60 missiles targeted at the United States, but they are a sophisticated type with multiple warheads.[10] A defense cannot stop that many missiles, so it comes down to how many are going to be launched and how accurate and survivable they might be. North Korea has none, so far as we know. They are at the "capability" stage; i.e., they can show that they are capable of launching a missile, possibly mated to a nuclear warhead, and one that may be able to hit its target. There are a number of "ifs" in that statement.

First, they have to have a supply of nuclear warheads, so that if they shoot some off there will be a few left over. Otherwise the consequences of the first strike will be severe, with no chance to retaliate. Second, they have to make the warhead light enough to put on a missile that they have in their inventory, light enough to be launched. Third, the missile guidance system has to be accurate enough to put it on a target they want and not just throw it off in the ocean or some cornfield. For North Korea, there are few intelligence guesses about whether they could do these things or not. Not so with China; the U.S. knows China could launch missiles at it.

That debate was not about whether North Korea could launch a missile a substantial distance, but whether there was urgency to do something about it. Why we are having the same discussions twenty years later is of concern. We built and fielded a National Missile Defense system, which was supposed to defend the U.S. from these types of attacks. It cost $40 billion and has 30 interceptors, most at Fort Greely, Alaska. There will soon be up to 44 interceptors in the ground there.[11] If National Missile Defense actually works, the North Koreans could launch quite a few missiles before the U.S. citizens would see incoming contrails.

China has enough weapons to show capability and have some left over. Our intelligence services probably know if they are mission ready and if they can hit their targets. The measure of accuracy is something called "circular error of probability" (CEP). The Russians, during the Cold War, were thought to have quite a few missiles that were range-capable but not very accurate. They had high numbers in the CEP. They made their warheads bigger to compensate for their lack of accuracy. They were big enough weapons to make the term "close enough" more relevant. Larger weapons require bigger rockets, which the Russians had. China does not seem to care about this aspect, suggesting they have more accurate missiles or are not concerned about having a few miss by a few miles.

The problem with estimates of capability was demonstrated after the Cold War was over: We were able to determine that the Russians did not have as many weapons or rockets as we had thought. Wheeler repeats some speculation that the Chinese might not either. Nobody knows very much at all about North Korea, so we could be wrong about them too.

But the Chinese narrative is something unrelated to the facts at hand. They have said that North Korea does all of its testing of weapons because of the threat that THAAD would be put in South Korea, not the reverse proposition that the missile testing in the North was the reason for THAAD's introduction in the South. The U.S. has been worried about North Korea since long before THAAD was even invented, and it seems they were right about their concerns. China has kept this narrative going for such a long time that we have forgotten they are behind it.

For the past few years, the United States was diverted from talking about China's actions to support North Korea or steal intellectual property from U.S. companies and use it to develop its own product lines. That shift came after China and Russia came to an agreement in 2015 that was intended to limit conflicts of various types between the two. There were a total of 32 bilateral agreements signed between them, and one in particular was a "nonaggression pact" in cyberspace, with a promise to avoid destabilization of politics via the Internet.[12] After that, the Chinese seemed to back off of hacking as they promised; however, they may have been leaving the unsavory parts of political wars to the Russians.

When the Russians took Crimea, a land mass of 10,000 square miles owned by Ukraine, they did it without firing a shot. Sun Tzu could relate to that. Nobody from Europe poked their noses into what was happening in the build-up of armed forces rolling into the region, although we know the Europeans would have had their own intelligence operations telling them what was about to occur. They know the Russians are in Ukraine in numbers that continue to fight. They know the Russian propaganda machine generates narratives that fit the Kremlin view of the world and plants stories on social media to confirm those views.[13] They know the Russians are not going to quit their push to take back territories that were part of the former Soviet empire. The Baltic states are under siege from the Russians every day, and the Russians bought television stations in Europe to further their influence over the views of people living there.[14] Their strategy seems to be working in many places.

Before Crimea, the Russians sent troops into the rest of Ukraine but denied having anyone there. This, in spite of a Facebook post by a Russian soldier with his unit, showing location information from both sides of the border. As I outlined in my previous book, *The New Cyberwar* (McFarland, 2015), the Russians criminalized the behavior of leaders in Ukraine, put up posters showing them in a disparaging light, directly campaigned against them, funded their opponents, and implied that the current government was a return to the Nazi era. They paid for billboards that showed half of Ukraine with a swastika laid over it and the other half covered in the Russian flag. They padded the election rolls with people moving from one polling place to another via bus, and tampered with the computers that assembled the completed voting tallies and counted them. This is much more than was done in the 2017 United States election.

In Ukraine, the Russian nationalists captured and turned over to Russia a female pilot, Nadiya Savchenko, who has done well for herself in captivity, being elected to the Ukrainian Parliament. She was traded for two soldiers whom the Russian Embassy called "Russian citizens detained in the Luhansk region," which might make them tourists, but definitely not soldiers. They claimed to be soldiers, making it all the more difficult for Russia.[15] They could have been shot as spies but for their confessions in which they called themselves "contract soldiers" who were there on leave from sabotage missions. One said he still had his contract with him. Every satellite could see them and where they came from, so it left little doubt that the Russians were invading Ukraine, however subtle and covert it was supposed to be.

All the interested parties know what is really happening because they spy on one another; they get information from friends of theirs that are better at it than they are. They trade that information like poker chips, and would have been prepared for what happened. Yet they do nothing, just as they did nothing when Hitler seized Poland in the name of annexation. Peace in that circumstance had a price, well known to the people of Europe, yet we seem to believe that the Russian seizure of Crimea was somehow different.

The E.U. and U.S. continue to do little except level sanctions on selected Russian businessmen and their banks. In December 2016, the U.S. added the names of thirty-seven individuals, plus seventeen Ukrainian separatists, and eighteen companies operating in Crimea, nearly two years after the invasion of Crimea occurred.[16] Did they really believe that the Russians were going to pull out of Crimea because of sanctions? Nobody on either side could believe that because it is not credible. This battle is over, and the Russians have won, but not one government official on any side called it war. If only temporarily, it ended by an accident of war.

On 17 July 2014, the rebels shot down a Malaysian commercial airliner with a Russian Buk anti-aircraft missile, crystalizing resentment against their cause. The Dutch Safety Board issued a report of the incident in October 2015, outlining the facts as they were stated in the West.[17] The Russians not only denied that the rebels had used Russian missiles; they also claimed Ukraine shot the airliner down, thinking it was Vladimir Putin returning home. The story was so preposterous that it got little traction outside Russia.

The Europeans and United States issued sanctions, and continue to enforce those sanctions, long after the Russians got the message that they need to be more careful with those weapons.

This battle is far from over. The Ukrainians, according to the Russian news service RT, sent covert forces into areas of Crimea to sabotage progress in the Kerch Strait, a partially submerged area between Crimea and Russia that Hitler considered as the path into Moscow to save his invasion. Someone certainly blew up electric generators on the Ukrainian side of the grid, and parts of Crimea had no power until they got their generators going. The Russians built an "electricity bridge" along the same route and are trying to build a real bridge from more solid ground in Russia because they are not able to supply Crimea without depending on Ukraine.[18] It turns out the costs are more than Russia can afford. They solidified their holdings, and prepare to take more, showing only that a successful strategy is one to be repeated. This kind of engagement keeps going until the achievements of the combatants are legitimatized, which for Crimea seems to have happened.

The closest thing we have to war is in Syria; though most of the fighting is done by countries that have not declared war, several call it war. It is fought in multiple countries with state and non-state groups that favor no one but themselves. Al Qaeda and ISIS, the best known, use the Internet to recruit, train and motivate their members, who come from countries all over the world, threatening some very powerful governments including Iran, Iraq, Syria, Libya, and Saudi Arabia. Participants, particularly ISIS, seize territory in a number of countries that object. Governments, as a rule, do not favor this kind of activity, and slowly, selectively strangle ISIS members in their own beds. This process drags on forever, but appears to have the desired effect.

There are so many conflicting stories of attacks, bombings and named groups involved in combat that it is impossible to discover which stories are true. The Russians are said to bomb a hospital in Aleppo for reasons nobody can figure out, then bomb hospitals again in April 2017. In Aleppo, they denied it but bombed it again to make sure they got their objective. American aircraft bomb Syrian troops in subsequent actions. This is the way lessons are learned in the new kind of war. You bomb our covert forces and we bomb yours. Nobody is at war, but the actions can lead to the death of innocents of all kinds.

The Kurds push ISIS out of territories they then occupy themselves; the Turks bomb the Kurds and shoot down a Russian jet over the southern border, but become friendly with Russia again within months after a coup tries to overthrow the Turkish government. The Turks blame a murky cleric who lives in the U.S. in the northeastern state of Pennsylvania, for planning the revolt. The U.S. is unwilling to turn him over to Turkey. The Russians cannot help the Turks with that, but are friendly nonetheless. As the friendship stabilizes, the Turks roll tanks into Syria and kill ISIS members and some allies of the U.S. who happen to be Kurds. Everyone bombs the Kurds. Though they are still among the best fighters in the world, seemingly no country loves them enough to help them with their building of a country.

We should remember that operations against ISIS are a so-called civil war that is being fought in Iraq, Libya, Tunisia, and a number of other places. We might notice the conspicuous absence of Iran in that regional view, even though we know Iran supports most of the groups acting for Syria's benefit.

A civil war is generally fought between citizens of the same country, though there may be few limitations on the geography. Nobody checks citizenship in this dispute, but the fighters seem to come from several different countries, many of them returning home after they get some training. This certainly looks like no war ever fought before. What we are seeing is a change in war, one that sounds like it favors fighters almost anywhere by having fewer causalities. No more charging down a hill to certain death. Now, we hide our dead with denials and deflection, but we are only fooling our citizens with poorly kept secrets. Eventually they find out.

The use of military and covert forces is a trick, a transparent deception, that populations have accepted. Most wars have become covert conflicts; i.e., neither side admits to being at war and keeps the threshold of conflict low. This means little to innocent bystanders who become refugees from their own homes. It means nothing to the participants who may be Russian soldiers who sign up for paid assignments while on leave from their regular units, or U.S. "advisors" who fight alongside their trainees in Afghanistan, Syria and Iraq. Jets bomb them just the same, and the other side shoots real bullets. As much as they want to hide and obscure the actions, everyone knows this is real war or some sort, even if it begs a definition. It is certainly *not* annexation.

What dominated the news during the U.S. election was the Obama administration's belief about a Russian plot to influence the election for president. The disclosures showed the inner workings of a political party. The content of those e-mails showed that the candidate had the questions that were going to be asked in a CNN debate that was supposed to be unbiased. The provider of that information was removed as a CNN contributing consultant. The disclosures showed the bias of the leader of the Democratic National Committee towards the election of Hillary Clinton, when the leader is supposed to be neutral and publically denied being biased. She resigned her position under a wave of unfavorable publicity prompted by the disclosures. The disclosures showed disdain for some minority groups whose votes were needed for the election. It showed potential criminal offenses associated with the destruction of government records. These releases, dribbled out from Wikileaks for weeks, set a tone that was not favorable to the Democrats, who eventually lost the election. We should not wonder if the Russians really did affect the outcome of the election, because those statements certainly did. But there may be quite a bit more to the idea that it did not affect the outcome as much as they would have liked.

In 2011, the Russian news service RT claimed that the United States had tried to interfere with the internal election in Russia while Hillary Clinton was secretary of state. RT published a lengthy outline of what it considered to be U.S. State Department interference with the Russian national election, starting with communication between the Hillary Clinton–like State Department (through USAID) and Golos (an independent election watchdog in Russia). They claimed persons who worked for State communicated with the Golos executive chief and deputy. The RT articles claim Golos was paid for violation reports, and before the investigation was finished in Russia, Hillary Clinton criticized the election results as "unfair." RT quoted Vladimir Putin as saying, "When financing comes to some domestic organizations which are supposedly national, but which in fact work on foreign money and perform to the music of a foreign state during electoral processes, we need to safeguard ourselves from this interference in our internal affairs and defend our sovereignty."[19] Dov Levin, writing for the *Washington Post*, described Vladimir Putin and Hillary Clinton's relationship as rocky—they were not friends.[20] Mrs. Clinton had a long record of criticizing Putin directly, and held the most hawkish views on Russia of any of the Obama foreign policy team. So if Putin really believed the U.S. was interfering with Russian election processes, and he believed that the author of some of those actions was the hawk in the White House meetings, he might have sought some reciprocal action. This is the nature of this small part of information war.

The year before the Russia-China agreements, Sony Pictures was hacked by the North Koreans, and following that, there was a release of e-mail between various parties, combined with threats to theaters promoting a movie critical of the North Korean leader. This resulted in the removal of the movie from many of the places it would show. Although it did not stop the show from being seen later, it reduced the public exposure and may have had an effect on its profitability. But this was a warning that was not directed at Sony. This was China's proxy, North Korea, warning the United States that war was changing and we would not like the exposure of similar information in the public.

At the same time, China stole the records of security clearances of many of our top government and business leaders from the Office of Personnel Management, emphasizing the point further. That was not discovered until 2015. That information includes arrest records, drug treatment, investigations by businesses and government offices, financial information, credit ratings, and foreign national relationships, among other very private things. Washington, D.C., where many people have things to hide, collectively shuddered. The release of that kind of information could prove more embarrassing than any of the e-mails from the DNC or Sony, yet none of it has been released— that we know of. Its value lies in not being released, but in having every person with a security clearance know the Chinese have it. It is influence, albeit indirect.

It is difficult to see disclosure of private information by these two countries as part of war, unless we see it in context. In information war, no buildings fall down when information is stolen. No lives are generally lost from words alone. No walls collapse when a "narrative" describes events in the way a government wants it to be seen, and not the way events actually happened. Strategically, this war is about changing or bolstering perceptions of different audiences towards dissimilar political systems, the same approach offered in the U.S. Joint Staff publication on information operations.

A minority of people targeted by governments believe in the press as their guiding light, and even fewer believe in politicians. We hope a free press will find its way to exposing the truth, but only a minority of people in the U.S., China and Russia trust the press as a credible source of news.[21] In the U.S., for example, 75 percent of those surveyed in 2016 thought news outlets were biased, yet those same media tended to keep politicians in line.[22] We can, of course, find polls that say the opposite. Yet as we came to find in the U.S. national election, we can hardly trust polls anymore, because they seem to show what sponsors and news outlets want to show. Similar polls predicted a landslide victory for Hillary Clinton. Media outlets in the U.S. are thought to be bought and paid for by business leaders with political alignments to one party or another. To some, the U.S. political system is not much different from its major rivals in the world where the press is tightly regulated and dissent is not tolerated.

The kind of management of information varies, but the root of the problem is that political groups can influence what news is presented and how it is slanted towards a narrative that the leadership believes is important. What the leaders fail to realize is that the public is much smarter, and more cynical, than the world leaders given them credit for.

The United States, China, Iran and Russia are far too similar in the way they conduct this warfare, though they have vastly different forms of government. It should be discomforting to know that governments use some of the same methods, but it should not be surprising, because those methods are effective. Because operations like the hacking of the DNC, the leading French candidate, and German candidates are more visible than Chinese hacking of embassies and political leaders, we think the Russians are meddling in almost every election. What they are showing by consistently getting caught is their clumsiness in their approach. They have demonstrated that the release of information shows some promise. Those actions will be duplicated in the future by more governments because manipulation of domestic populations and their political leaders is the major aspect of winning a battle in this kind of war. The subtle difference is that the Chinese do it without the level of exposure the Russians have had.

Governments prefer the use of persuasion through the control of information that populations receive. They manage their perception of events and influence what different audiences believe about government actions. Press controls are more common than we generally think about, yet few countries control the press better than Russia and China. In its 2016 annual report, Freedom House, an independent organization monitoring freedom in several forms, says only about half the countries of world have a free press, and the number is *decreasing* as more and more groups try to use the press as an instrument of their politics. Particularly in the Middle East, pressure to report stories consistent with government views has reduced the ability of the press corps to publish stories that are out of bounds. Some, such as Egypt, Iran and Turkey, have become more aggressive in controlling the major news outlets.[23] Governments do this around a narrative that weaves certain facts together in a way that creates a belief; they can discourage counter-narratives and encourage parallel views. Different governments use techniques that vary but produce the same result.

The Chinese ban certain stories that do not fit the narrative and enforce those bans with censorship. The ends to which censors go to do so are remarkable. In February 2017, the BBC interviewed a woman colloquially known as "the Kung Fu Grandma," a practitioner of the art at the age of 94.[24] She was still independent and explained Kung Fu as a method of protection from potential bad actors. The BBC was accompanied by a few local government officials who outnumbered the BBC crew. Their purpose was apparent as the interview went on. The subject of her religious beliefs was about to be discussed by family members when they were stopped by the officials. Religion was out of bounds. Nobody can rationally explain why religion is not an allowable topic of discussion for a woman who is that age, but explanations are never given for censorship. The Chinese go to extraordinary lengths to limit what the press reports about any, even the most trivial, news stories.

The Russians use the RT news service to create and support its narratives. The press is more crudely managed in Russia, and the well-known persecution and execution of some press representatives makes sure the news is reported in the way the Kremlin wants. Independent news services press the limits of that control and often get stories into print that conflict with the official view, but being in that business has been dangerous. Fewer reporters end up dead, but more of them have resigned "for family reasons" that at any time in the past.[25]

The U.S. has a massive press corps responsive to political parties and special interests of their own making. Government involvement in managing the press is largely done by influencing friendly outlets and favoring them with stories that are consistent with the government position. Those reporters in favor get access to politicians and staff who provide current, credible information.

But in each of these cases, it is far more complicated than just managing news media, because it means contrary narratives must be prevented or discredited. In many cases truth does not matter, but enough truth must be present to make a narrative credible. The biggest battles come when the narratives of countries are at cross purposes with each other. We can all remember the example of the commercial airliner shot down in Ukraine. Narratives abounded for almost two years before we saw an authoritative report blaming the rebels in Ukraine and pointing to a missile of Russian manufacture as doing the deed. By the time that account was published, only a small number of people could remember the event that had caused it to be written.

Control of the press, censorship of various kinds, and political movements of military forces to meet political ends are only a small part of the larger effort to stake out territory, seize and hold it against the wishes of the affected countries and the United Nations. The construct that describes how this was done is not new or even very clever. Thirty years ago, it was called information war, but it existed long before computers came along.

4

Political Wars

A year before the start of World War II we might have asked ourselves why we should worry about war with Germany, Russia, Italy and Japan. Russia was a relatively new communist country, not ready for war, and it had signed a peace treaty with Germany just days before the invasion of Poland. Italy was hardly a world power, but it shared some common interests with the Nazis, especially its concern about communism in Russia. Japan was a world power and had plowed through large parts of China that it retained; it had dominion over Taiwan, and it controlled territories given to it after World War I in Tsingtao, on the Chinese Shantung Peninsula, and on the formerly German islands in Micronesia. The United States had disagreements with Japan because Japan held territories captured during its war with China, but the people of the U.S. were not thinking about war with Japan. Germany was of concern to Europe and Japan but not to the U.S. until much later.

What triggered the events that led to World War II, in a simplified portrayal, was a series of events that came from expansion of territory that became intolerable to those offering appeasement to Germany and Japan. These were not the only causes, but they were significant enough that Europe finally had to admit that appeasement was not working and it was time to block further expansion. Japan was at odds with the League of Nations over its territory in Manchuria. The league said Japan did not have the right to seize the territory. Japan ignored any attempts to be removed.

Japan was working for a diplomatic solution that would satisfy the United States, but those attempts were not well received. President Roosevelt and his secretary of state, Cordell Hull, were almost antagonistic towards the overtures made by Japan; the conflict culminated in a series of export restrictions on commerce.[1] After that, the diplomats on both sides realized that war was a real possibility.

So, who in 1937 could have predicted that Germany, Italy, and Japan would be allies in a war against Europe, England, the U.S. and their allies—a true world war? Not many people did. Not many people today see a war coming either, and those who do will not want to say it out loud. It is not popular, and governments do not like to debate theoretical events. However, they are no more theoretical today than they were in the weeks before World War II began for Europe.

The groundwork for political warfare solidified in those first weeks of World War II in England's Political Warfare Executive, which operated radio stations that appeared to be German stations but were not. The U.S. operations were much slower in developing and concentrated more on deception in military operations. Neither country's operations were very successful.[2] Both became more successful after World War II by focusing on delivery of news—the BBC World Service and the U.S. Voice of America. These had the ring of a popular belief that the truth will make us free. Presenting the truth would allow the West to preserve its values in the countries of Europe. We have never stopped believing in this sentiment, but we are finding the truth more difficult to define and support.

Each country uses information as a weapon of influence and persuasion, both for its own people and for the part of the world that chooses to try to bring new ideas to its people. Weaponized information blocks and manipulates anything or anyone that does not have a view consistent with that of the central government, including those outside their own boundaries. Most countries control the press and media to achieve that purpose. Democracies and totalitarian states look more alike when they both conduct the same types of operations.

The techniques countries employ are part of a larger kind of information war called political warfare. While information war was a military creation, political warfare was not part of it, thus not mentioned in that doctrine. Political warfare in modern times was defined by George Kennan as "the logical application of Clausewitz's doctrine in time of peace. In broadest definition, political warfare is the employment of all the means at a nation's command, short of war, to achieve its national objectives. Such operations are both overt and covert. They range from such overt actions as political alliances, economic measures (as ERP—the Marshall Plan), and 'white' propaganda to such covert operations as clandestine support of 'friendly' foreign elements, 'black' psychological warfare and even encouragement of underground resistance in hostile states."[3] It is difficult to see these kinds of activities because they are covert, hidden both from the public and from most of the governments that carry them out, but this is a description of war today.

Even the United States considered the practice of political warfare against the Soviet Union in the Cold War and developed a report on the overall problems with conducting it.[4] We can remember the Star Wars initiative of President Reagan, which framed a defense against missiles, yet the physical system that was proposed was never built. That defense would have undermined the mutually assured destruction strategy that allowed the Russians and Chinese to maintain small missile forces with nuclear warheads while the United States, France and the United Kingdom had delivery systems of their own. With the destructive power of offensive systems dispersed across wide geographic regions, the introduction of a defense against strategic missiles was big news. Russia could not maintain the pace of countering the military objectives of Star Wars and the attempt eventually weakened the country. China did not get involved, but it evolved a strategy that was not about disseminating truth as much as redefining it.

Inside China, the central government has historically punished views that differ from official policy. But over the last ten years, the government has developed the capability to filter the content of social and news media, and it stifles institutions including religious organizations, businesses, and any other group that has a message that differs from its own. What makes China the world leader in political warfare is that it puts time and resources into doing things to control and manage the content of any message that is conveyed to the people in ways that no other country can or will. It takes organization, a central purpose, and a good understanding of human nature to make that work. The Chinese act as though they believe they can convince the world to accept their ideas by managing information about those ideas. We should look at their success before saying they are wrong.

It seems China has no enemies. That idea is discouraged by press, politicians and business leaders because it is necessary to economic progress, the central part of information war. If we want to have a clear conscience, we cannot trade with our enemies, or at the extreme, exchange more than negotiating pleasantries with them. Our businesses cannot operate their production facilities in the enemy lands, nor can their laborers keep those facilities operating. We would find it difficult to sell or buy goods of an enemy. The globalization of trade, particularly Chinese trade, has almost eliminated enemies. Without enemies there can be no wars.

This is hardly a new phenomenon. In the postwar era of the late 1940s, those soul-searching in the United States examined those businesses and governments that had cooperated with Nazi Germany and Japan, which were clearly enemies. For the most part, that examination was not very conclusive and was clouded by denials from every company named, even failing to distinguish between opportunism and outright cooperation. Except for some parts of the insurance industry, very few calls for retribution were ever addressed. Any cooperation with the Axis countries was discounted or ignored.

In the post–World War II era we did not dare to call a company's behavior traitorous when it traded with an enemy in time of war, so there is no reason to think that we would do it when we are not at war. These days, there are no wars, and except for violations of export laws, trade is unlimited. That is view that heavily favors the manufacturer of goods and services. Globalization makes war impossible.

Even where war stares us right in the face, there is a reluctance to characterize the circumstances as war. The U.S. Congress is reluctant to declare war against ISIS because of where it could lead. Once such a declaration is made, as it was in Vietnam's case, the president has wide latitude on how to carry out that war. For political reasons, Congress does not want that kind of authority to be ceded to the president to deal with terrorists who have no country and might operate anywhere. So, in spite of frequent pronouncements about war with ISIS, the U.S. is not at war with them. We fly airplanes that attack ISIS; we have soldiers fighting alongside troops who engage them; we cut off the flow of money to them; and we treat them like an enemy. But the one thing we do not do is officially call our activities war. If this sounds inconsistent, it is because it is.

But what the Chinese have done with war is far more advanced. This is a People's war, the body and soul of China. The population cooperates and the soul, the Communist Party, directs. China combines its state government, its businesses, and its military into a shaped charge that penetrates any obstacle that opposes it. Mostly, that means directing its activities towards the biggest and most powerful competition it has, the United States. China's approach is a relatively new phenomenon, but it is changing fast, and it is spreading to other countries. Like Crimea's seizure by the Russians and the rise of ISIS, this conflict takes place below an imaginary threshold that we define as war.

As a simple example, the U.S. defines a cyberattack against our electricity grid as an act of war.[5] But when somebody used a cyberattack to take out parts of the Ukrainian grid, leaving 230,000 people without power for six hours, it was not called an act of war. The U.S. Department of Energy claims Russia did this attack, indicating that the combined investigation done by the Energy, State and Homeland Security departments and the FBI came to that conclusion. But the intelligence community indicated it was too soon to draw such a conclusion.[6] The attack comes from Russia, and we would not like to pretend the Russians are at war with Ukraine. They are simply annexing territory. The Russians pretend they are not involved there, in spite of considerable evidence to the contrary. The attacks are grown slowly, over a period of months, using prepared code called BlackEnergy3, which was deployed in Europe and the U.S. The Russians launch denial-of-service attacks against call centers of the victims who might report power outages, and erase disks in the distribution centers to get rid of the evidence of an attack.[7] This is a low level of warfare of the kind that usually precedes a more aggressive attack—which did not follow. Perhaps the next time, it will, or this could just be a warning to anyone thinking about helping out the Ukrainians.

The governments of countries in Europe, North America, and Asia are skeptical—some even leveling sanctions—but are not convinced that Russia and Ukraine are at war. They attribute the electrical infrastructure attack to groups of hackers who have no direct association with the Russian Federation. The use of third parties clouds the matter of who is to blame, but most governments know who is really involved; their intelligence services tell them. But of course they cannot make that information public, because it is a state secret. That is an excuse that works for many countries, not just the Russians and Chinese.

The North Koreans attack major parts of the banking infrastructure, businesses, and government offices of South Korea in a destructive attack that is not war. They sink a military vessel, killing members of the crew, something clearly an act of war but ignored by all. North Korea or China attacks Sony in the U.S. over a movie being distributed and shown there, and that is not war, but a warning of what war could look like if it happened. This is an information war, where—at the lowest level—information originated by someone in a private e-mail can be seen by anyone. That release is a political tool used by cyber groups like Anonymous and political ones like Judicial Watch, but when governments sponsor and carry out the attacks it is more than that. It is threatening U.S. business leaders and government employees who have much to fear from this kind of attack. What the Chinese have done is build a deterrent strategy that warns of the ability of cyberwar to influence and enable political objectives. It is a form of veiled blackmail. This is, of course, the nature of deterrence. To be effective it has to be credible, and it seems to be easy to take a step from where they are now to making some of this information public if that will support their cause.

In the same vein, there are the missile tests done by the North Koreans. Live tests of nuclear weapons are only done by rogue states, because other countries with nuclear weapons have signed the nuclear test ban treaty to keep the number of tests down. Shooting off missiles that can carry nuclear weapons long distances is equally disrupting. Making public statements that the purpose behind testing nuclear weapons and missiles is to make preparation for a war where they will be used in combat with the United States is mostly bluster, but still not something to ignore. Bruce Klinger at the Heritage Foundation described China's willingness to support North Korea in spite of its transgressions as a series of steps that allow them to continue their antics:

- Repeatedly resisting stronger sanctions,
- Watering down proposed [U.N.] resolution text,
- Insisting on expansive loopholes,
- Denying evidence of North Korea violations,
- Blocking North Korean entities from being put onto the sanctions list, and
- Minimally enforcing resolutions.[8]

But it is not just the information war that allows China and North Korea to win the first round in this engagement. This is just one strategic element of a broader buildup of arms and territory. Of the most concern today is a build-up of islands in the Spratly and Paracel island chains, where runways as long as the ones at Dulles airport in the suburbs of Washington, D.C., have been laid on areas that once were below water. They created the islands from spots of land so small that nobody lived on them.

To those who might question the validity of their claims to those places, China can say, "Those islands were just created by us from spaces long claimed by us." Those spaces they talk about are half the size (1.4 million square miles) of mainland China. The U.S. believes the claims are dubious, and not recognized by very many other countries, but that will not deter the Chinese.

Vietnam, Japan, Indonesia and the Philippines would agree, but it was the Philippines that actually brought an action in the U.N. Court of Arbitration for the settlement of these kinds of disputes.[9] China did not attend the proceedings, which had the principals speaking to themselves, but speaking nonetheless. The Philippines claims the whole tussle is about China's original claim to territory occupied by a retreating Chiang Kai-shek before the Communists could overrun him on the mainland. If we think about that, China is, as it has always done, claiming Taiwan as its own territory. In 1992, representatives of China and Taiwan actually came to an unusual agreement to allow China to claim the territory as its own, but to allow Taiwan to publicly speak about a different interpretation. The Chinese call this the 1992 Consensus, suggesting it is a government/international consensus on Taiwan. In May 2016, the Chinese criticized the newly elected Tsai Ing-wen for not including language about the 1992 Consensus in her opening speech. The Chinese cite U.N. Resolution 2758 as making China the only representative to the U.N. and the 1992 Consensus as justification for pursuing the strategy to make one China, with Taiwan and other land masses in the South China

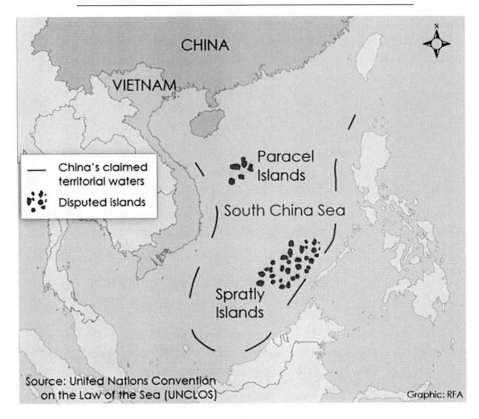

Figure 2. The nine-dash line (United Nations Convention on the Law of the Sea).

Sea included in it.[10] There is a constant drum beat of one China, even to the point that China criticized the newly elected president for accepting a call from Tsai when she congratulated him on winning. They succeeded in getting the then-new President Trump in the U.S. to back a claim that the "one China policy" might be negotiable. China becomes the "nagger," constantly ready to jump on any suggestion that there is more than one China, while wrapping Taiwan into that statement and moving to control the physical space around the island. We accept this logic without ever realizing the fallacy of it.

By raising the level of control in areas that surround Taiwan, China will eventually control the islands themselves. Like Russia in Crimea, the takeover will lack the drama of a real war, only because it is kind of bloodless war. China will use force, persuasion, and repeated statements of their positions over long periods of time. They wear down and intimidate their opponents, sometimes with more than words.

In February of 2016, China put a surface-to-air missile battery on Woody

Island, not far from Chinese airspace and a place claimed by Vietnam. When the effect was over, they withdrew the missiles. In March of 2017 they introduced J-11 military fighters. This ratcheted up the risk of flying over that territory. After Turkey shot down a Russian aircraft flying in and out of Syria on bombing runs, the ability to interpret what the Chinese did as a routine defense of their airspace may raise some military eyebrows. Missiles are there for a purpose, or at least the threat of that purpose. Modern Chinese fighters and other military aircraft have been in and out of that same island. That missile system is now being augmented by high frequency radar.[11] The Asian Maritime Transparency Initiative at Washington's Center for Strategic and International Studies said the images showed that construction of facilities was nearly complete and would allow monitoring of many trade routes in that part of the world.[12]

The Chinese respond to criticism of their actions with a simple, often repeated phrase: *We are not doing anything that any country would not do to protect its territory.* This linguistic trick ignores the disagreement over exactly who really owns this territory, and appears to state an obvious fact agreeable to almost everyone listening. This kind of logical manipulation is not uncommon, but the facts of the settlement of small chains of islands, in the middle of nowhere, are not quite that simple.

In April 2016, the G7 Summit that was taking place in Hiroshima, Japan, stated, "We are concerned about the situation in the East and South China Seas, and emphasize the fundamental importance of peaceful management and settlement of disputes. We express our strong opposition to any intimidating, coercive or provocative unilateral actions that could alter the status quo and increase tensions." The Chinese government thought this statement was aimed directly their way, and we can be sure that was a correct assessment. It complained, saying, "We urge G7 members to abide by their promise of not taking sides on territorial disputes, respect the efforts by regional countries, stop all irresponsible words and actions, and make constructive contribution to regional peace and stability."[13]

The island building in the South China Sea is the political and military strategy playing out in a contentious geographical area; the events would seem to be a matter of dispute over territory that has oil under all that water. But there is more to it than just oil. The South China Sea is one of the most heavily travelled trade routes in all the world, with over $5 trillion worth of international shipping passing through it each year, mostly from China to other countries. The Chinese would like to have that area under their control, but the oil and gas are a bonus. The island building program is just one way of taking over something much bigger.

A related initiative is to try to negate the trade agreements the South China Sea countries have with the U.S. and each other, because China sees economic warfare, a major component of information war, as an important part of engagement with the United States. In 2014, China's foreign minister promoted a national initiative to consolidate the number of free trade agreements to "reduce the risk of overlap and fragmentation," without really saying what benefit there might be. The idea was first raised in 2006 as a means to put China's economy at the center of trade in that region, a concept called "community of shared destiny."[14] The countries of that region seem to have little interest in pursuing such an arrangement, and have not even shown an interest even in conducting a feasibility study. As a simple matter, it was a way to reduce the influence of the U.S. in the South China Sea, and most of those countries saw it for what it was.

The island building is about to come to a new phase. During the Obama administration the U.S. sent an aircraft carrier, two cruisers, and two destroyers to join two others already there. The Trump administration has done much the same with different ships. While the ocean is large enough to accommodate them, they constitute a formidable military force that creates a risk of clashes between Chinese and U.S. forces. The Chinese have warned that the "freedom of navigation" shows of force are fraught with potential risk of clashes. What will bring it to a head is something called an air defense identification zone (ADIZ).

In 2013, the Chinese started to enforce an ADIZ in the East China Sea (northeast of Taiwan), and according to Philippine justice Antonio Carpio, they are beginning to enforce the same kind of zone in the South China Sea.[15] In 2015, the BBC hired a small plane to fly close to one of those islands, and the warnings they received were clear and persistent. Listening to them while the air crew looked nervously around left no doubt that the warnings could be more than just idle talk.

The U.S. military has already publicly announced that it will not recognize the ADIZ in the South China Sea.[16] The implication is that U.S. military jets will fly into those warning areas at times when Chinese air defense missile systems are deployed on the islands. That should make for some interesting engagements. Pilots of those aircraft will know the radar from that missile system has locked on, and what they do next will determine how safe that area will be moments afterwards. We are going to see some concrete tests of how both national strategies play out.

The Chinese may also have been amused to see how quickly the competition for islands increased when the U.S. made its moves there. Vietnam, Indonesia, and the Philippines quickly seized additional fishing vessels in waters claimed as sovereign territory by more than one country, something they have done individually for some time.

China must have known the U.S. would do something about their continuous buildup of the islands in that area, and they were probably not surprised at the number of ships that were coming. They have hacked into our military networks for years and have a good idea of what military forces are up to. China combines its political and intelligence targeting in a symbiotic way. From a strategic standpoint, control of that space is reason enough for an aggressive country to want to claim it. The Chinese describe that country as the U.S., and the U.S. portrays it as China. But the are holds significance beyond a simple debate; its assets include these:

- Fish stocks and hydrocarbons. The ECS and SCS contain significant fishing grounds and potentially significant oil and gas exploration areas. Fishing boats are a common target of the Chinese and governments that dispute their claims in the area.
- Military position. Some of the disputed land features are being used, or in the future might be used, as bases and support locations for military and law enforcement (e.g., coast guard) forces, which is something countries might do not only to improve their ability to assert and defend their maritime territorial claims and their commercial activities in surrounding waters, but for other reasons as well, such as improving their ability to monitor and respond to activities on or near the mainland areas of other countries in the region.
- Nationalism. The maritime territorial claims have become matters of often intense nationalistic pride.[17]

Five years ago, the Chinese were harassing ships in the area, including those of the U.S. Navy. China's actions against intrusions are becoming more aggressive and persistent. The U.S.–China Economic and Security Review Commission 2015 Report to Congress says:

> Publicizing U.S. naval patrols and surveillance flights near China's reclaimed land features in the South China Sea appears to be part of a growing effort by the United States both to impose reputational costs on China and to reassure allies, partners and friends in the region as China's land reclamation and construction continue.... U.S. pressure on China to cease further land reclamation and military facilities construction appears to have been largely ineffective.[18]

But before they did reclamation or construction, China hacked the computers of key leaders of the governments with competing claims.[19] They wanted to have access to their positions, intentions, and reactions to Chinese movements. But for a little luck and good investigative work by the cyber security community, they might never have been discovered. This gives them the capability to know what another country will do at the same time that country is making the decision. We can almost bet that they have done the same thing in the U.S., before putting surface-to-air missiles out in the middle of nowhere. This is the role of intelligence collection in the governments of the world, and the Chinese do it well. It allows them make better predictions about what the other claimants to territory will do, and anticipate what their allies, like the U.S., will try to do about it. They follow the same model repeatedly, while others learn from them.

I mention this only because the Chinese have stopped playing our games of war and have started playing one of their own. They fight an invisible war with the United States, one where a battle is over before we realize it has started. Our military is at a disadvantage in its application because it is part of a system of government that is democratic, decentralized, and separates the government from commercial business. The Chinese are not democratic, are centralized, and combine commercial and government operations into one centrally managed state. That is an advantage that allows them to focus their limited resources in areas where they want success.

For the U.S. these last five years were a time of awakening, but it was reacting to more than just the claim of islands out in the middle of nowhere. The U.S. recognized that the control of space in the South China Sea, and the incessant cyber stealing from businesses, government, and individuals, is related. These are two parts of the same war. Recognition of how that part of a war is being conducted is not as easy to see as a rain of bullets carving up your companions walking down that hill at Antietam.

5

Fancy Bears, Chinese
Businessmen and U.S. Politics

Foreign involvement in U.S. elections has been an issue for many years, and each time the subject is raised, the press portrays it as a new phenomenon. This time though, in what is arguably the most heavily touted display of government force in an election, the Russians have hacked into mail servers of the Democratic Party political apparatus, engaging all the headlines by releasing what they found. The director of the National Security Agency said that series of actions did not have as great an impact on the election as those who did it thought it might. It is a twisted kind of comfort that it could have been much worse. Nobody was paying attention to what the Chinese were doing in this election, and they did not factor into the assessment of the effects.

Hacking is a term that is slightly overused, but in almost every public definition there is reference to the gaining of information through unauthorized access to computers that house that data. There are several other reasons for unauthorized access to computers, and each of these involves hacking of a different kind. Governments sometimes do not want information from systems they hack; they want to deny access to information for those who should have it; and they may want to alter the information so an authorized user receives data that is not correct. On a deeper level, they may want to put software into a computer to cause that computer to do something it is not intended to do, possibly at some point in the future. Each of those things involves a different kind of hacking, often operating under different authorities. We almost always think of hackers as unauthorized persons acting on their own, when they often are not. Some have government backers who give them immunity for their actions. Some have criminal backers who do the same thing in a different way. The techniques for entry are often the same regardless of the motivation, but hacking is different in each of those instances. Of all the things they do, stealing information is the easiest part.

Stealing information is so easy that anyone who wants to can buy the software to get them into another network and take what they want. It takes practice and the willingness to discover how to do it, but it is easier than when young people did it ten years ago. That is not new—hacking has always been ahead of our ability to defend large networks from people trying to get in. But the difference now is that the Dark Web will sell most anything needed to hack the best security in networks commonly used to grant users access to the Internet.

Russian and Chinese groups have taken that one step further, infiltrating sites that store software used in a number of commercial devices and installing their own versions of operating systems and applications. That software has been modified to give them access to or control of the networks that we use. They use software that is "verified" with stolen software certificates that look valid to security modules in other computer devices. The users cannot distinguish the bogus software from the authentic software posted by a vendor. That application downloaded from an external website may be the same one used on other computers, or it may not be. Service providers, with the possible exception of Google, do not provide us much protection against these kinds of threats, even if they have professional security staff working on the problem. Private networks, secured by political associates, stand no chance against the state-sponsored hackers who are good at what they do.

The ease of hacking makes it more difficult for hackers when governments go looking for people getting into their computers. Governments have, or can hire, technical services that can find groups getting in and identify how they did it. They often find two, working for the same government attacking the same target, apparently not knowing that the other was doing the same thing.[1]

That is the public story being told, a simple story that is easy to digest. This alone makes it suspect. A better story appeared in Reuters that describes a Russian think tank controlled by Vladimir Putin developing a plan to disrupt the U.S. election and discredit the electoral system in the United States.[2] The plan was said to have been prepared by the former members of the Russian intelligence services who make up the think tank. The plan was supposed to be helping elect someone more friendly to the Russians than President Obama or Hillary Clinton. Hillary Clinton was not very friendly towards anyone in the Kremlin.

If there was such a plan, it certainly did not disrupt the election or discredit the electoral system. That part is being accomplished with a campaign to discredit the winner and undermine the presidency, willingly helped along by some not-so-unbiased people in political parties. The main goal is to prevent the winner from governing. The Russians do not stop when the election is over. Sometimes their candidates lose; sometimes they don't turn out to be as friendly as they thought they might be; but they remain undeterred by setbacks. They have a long history of meddling, but they are much better at it than they used to be.

The target, in this case, was the Democratic National Committee (DNC), the leadership of the Democratic Party in the United States. The attackers have code names, Fancy Bear and Cozy Bear, given to them by the security community doing investigations of their activities. The names allow security groups to talk about techniques being used and where the groups usually live.

In 2014, a long time before the buildup to the 2016 election, Mandiant (later bought by FireEye, Inc.) gave the group doing the hacking of government offices and intelligence sources overseas a name: Advanced Persistent Threat (APT) 28, speculating that it was hacking for the Russian government and not for either China or international criminals.[3] The latter part was important because China and criminals were a hot topic then, and Russia was not. While Mandiant was looking for things Russian, they found another group, "among the most capable groups that we track," and put some staff working on finding out what this group was doing.[4] They called that group APT29. These were called Fancy Bear and Cozy Bear by the security community tracking them. A summary of the attacks on the DNC said Fancy Bear was caught by security groups, albeit after the fact of the first theft occurring; i.e., Cozy Bear was caught after Fancy Bear had been discovered. Cozy Bear had been inside the network for over a year, quietly collecting things typically used in intelligence collection but not by groups trying to make a living out of it.[5]

Cozy Bear acted like any other intelligence collection operation would. The group went in quietly and never disclosed anything outside of intelligence circles about what it collected. It appears Fancy Bear was after things it could publish and Cozy Bear was after things it could use to help decision makers in Russia, without anyone knowing the information was stolen. The disclosure of stolen information is seldom the way intelligence services operate, but Fancy Bear did not follow the usual rules; it was getting unflattering information out to the world public, clearly indicating where it came from. They used press outlets and Wikileaks, a popular website.

The techniques the Bears used were described by a man with some experience of Russian attacks on his country, Toomas Hendrik Ilves, the president of Estonia from 2006–2016:

> What we are seeing in the United States and among the European allies is that influencing a country's election outcome is warfare. There is no need to wage a kinetic war

or even use debilitating cyber attacks on critical infrastructure if you can sway an election to elect a candidate or a party friendly to your interests or to defeat a candidate you don't like. This is clearly the goal of Russia in the German elections, where Angela Merkel's role in maintaining EU sanctions against Russia has been critical and annoys Russia no end. It is true as well as in France, where Marine le Pen's Front National is anti–EU, anti–NATO and anti[–]US. With anti–EU and anti–NATO parties rising in popularity in a number of countries in Europe, this asymmetrical attack on the democratic process is already now a security threat to the NATO alliance.[6]

Ilves is giving some context to an already difficult subject that is generally not taken in context. He is saying that this is a broader war against democratic institutions being waged by Russia, when the U.S. seems to be fixated on what is claimed to be a Russian attack on the United States. It clearly is not. It is an attack on an institution, the election of officials in a free election, who govern somewhat like the electorate believes they should. It is an idea shared by many other countries, but not one found in Russia or China. Even though both would like to believe their elections are free and open, there is considerable evidence to the contrary. Hillary Clinton, while secretary of state, managed to insert herself in this issue and grab the attention of Vladimir Putin.

If both groups attacking the U.S. Democrats were Russian, it seems odd that they were not on the same page. Intelligence is a costly business, and having two agencies working on the same place at the same time is not terribly efficient. Having them work at odds with each other seems even more unlikely. Ilves says both of these groups are working for the GRU. For the GRU this was hardly breaking new ground, but it was obvious that the two groups were not working together. We have to think about this for a minute to realize that the Russian military intelligence was carrying out a campaign to disrupt the U.S. election. The GRU focuses on things that relate to the militaries of other countries. It would be as if the Defense Intelligence Agency in the United States launched a military campaign to influence the elections in Indonesia. Unless the military is out of control, that is unlikely to happen.

More than likely, it was the discovery of Fancy Bear that caused Cozy Bear to change its approach and forget about keeping secret the information it stole. After discovery of their operation, there was little point in protecting what it took. That speaks to its having been some agency other than the GRU. It sounds like the kind of thing the Russian FSB would do. There was other evidence that points to that, coming from the Obama White House.

The White House press secretary announced that the whole operation was directed by Vladimir Putin. That was supposed to be proof that the election was being undermined by the Russians so they could favor the Republicans. That announcement was not wise, and the information most certainly was very closely held in the intelligence community that told the leadership that it had occurred. Anytime a foreign intelligence service knows about such an announcement, it starts looking for who might have provided that kind of information to the United States. It did not take long to find some suspects.

In December 2016 and February 2017, stories surfaced about the arrest of two FSB officers and the leader of a Kaspersky Laboratory facility. The latter story by Radio Free Europe said two suspects were charged with treason, i.e., giving state secrets to a U.S. intelligence service.[7] These two officers could die because of the White House disclosure to the press.

Almost nobody willing to speculate about who actually gave what was collected to Wikileaks for publication on the Internet, and Wikileaks is protecting its sources. Since we only know that two groups of Russian hackers were discovered, there could have been other hackers who got the same information. Both of these groups would have known how to get something into print without showing the origin of the information. They have both run similar campaigns before, and their predecessors in the KGB trained them well. In the years before, they were involved in hacking of the White House, the State Department and the Joint Chiefs of Staff.[8] We would soon learn the FSB was behind the largest theft of credentials any company has ever seen when they stole nearly a half a billion user accounts in Yahoo.[9] The Russian intelligence services were busy.

The original release of the Democratic Party information did not come from the FSB or the GRU, as we would expect. It came from a hacker called Guccifer 2.0, which was a hacker pseudonym (most hackers use one to protect their real identity). Guccifer 2.0 claimed to have hacked the DNC. The name Guccifer comes from a real hacker, Marcel Lehel Lazar, arrested in Romania in January 2014 for hacking politicians and other government officials in the United States. The U.S. asked for extradition, and in 2016 he was convicted for hacking offenses and fraud. While in jail he told Fox News that he had repeatedly hacked the e-mail server of Hillary Clinton, something she and the State Department denied.[10] The curious gathered to see what there was to the story, but not much else came from the statements made to Fox.

Then, when the hacking was discovered and the information started to be public, the content showed a number of things going on inside the Democratic National Committee office. First, that the leader of the DNC, Debbie Wasserman Shultz, had been working with the Hillary Clinton campaign to sabotage the competition with her chief rival, Bernie Sanders, when she had previously denied that any such thing had happened.[11] Shultz resigned abruptly on the eve of the Democratic National Convention, where Hillary Clinton was nominated as the candidate.

Guccifer 2.0 started releasing more documents. Reporters, using their government sources, started to use a new term labeling Guccifer a "persona," which means something to intelligence agencies and hacker communities. It means Guccifer might be something more than a person; he might be an identity being used by more than one person. There might be multiple personas being used by many of the groups. There could be multiple people using those personas. We will never know all of those involved in making the stolen information public, nor the real identifies of people behind those names. The whole purpose behind a persona is to keep a secret—the identity of the real person(s) using it.

Lists of donations, Democratic donors, and the names, private phone numbers and e-mail addresses of these sensitive party assets were posted online.[12] These are the most sensitive secrets for any political party anywhere in the world. The Democrats blamed candidate Donald Trump even though the attacks started long before he was ever known as a candidate, and said the released documents might be forged. The Russians have long been known for faking documents and distributing them as legitimate, so the claim was easy to believe and fit the narrative that the Russians were trying to influence the election. Security groups brought in to investigate the hacks said they were linked to Russia, and the Obama administration sources told reporters they had confidence that it was the Russians. At the time, that seemed conclusive.

In October 2016, Guccifer 2.0 released a large number of documents though Wikileaks, which he claimed came from the DNC, the Democratic Congressional Campaign Committee (DCCC) and the Virginia Democratic Party, claiming they were stolen from the Clinton Foundation. The Clinton Foundation said it was not the target of this hack, and subsequent examination of their systems seemed to indicate that was true.[13] One of the claims made by Guccifer 2.0 was a pay-for-play connection of the Clinton Foundation and many foreign donors while Hillary Clinton was working at the State Department.

The release was the month before the U.S. elections and left little time to deny or present material facts on either side. The French elections had a similar event just days before the election. As we often observed in government computer security, in hacks of government and commercial offices, nothing involving attribution to a foreign government happens quickly. It takes time to identify which country actually did the deed, even where one country admits to doing it. That delay left the issue of attribution hanging (in this case, a public acknowledgment that the Russian government was involved in the hacking), leaving the matter to speculation on both political sides until after the election was over.

The number of releases, the quantity of internal correspondence released, and the issues those documents pointed to were disruptive and could have weighed on the perceptions of the candidates in the national election. But the majority of polls taken after the election indicate that the director of the National Security Agency was correct when he said the hacking did not affect the U.S. election as much as "the nation state" would have hoped.[14] In late December 2016, the FBI, Department of Homeland Security and the Director of National Intelligence issued a disjointed and hastily assembled report indicating the Russians were thought to be the attackers. This created a brief opening for politicians to claim that the presidential election was not legitimate and/or that the president was not a legitimate one. Protests and demonstrations against various policies continued for many weeks, hoping to gain some momentum that never materialized.

If this really was a campaign by the Russian government to disrupt the U.S. elections, it was not, if looked at narrowly, up to their usual standards. A Russian campaign would not have ended with the announcement of the winner of the U.S. national election. As the Russians did in Ukraine where they did try to influence the election, their actions go on until they accomplish what they wanted. In Ukraine, they did not stop when the election was over. They discredited businesses and personalities of elected and appointed officials and continue to do so today. If the Russians launched this campaign, it is not over. Their purpose is to disrupt the free elections of officials and constrain the government's ability to rule. Someone is doing that, but it may not be only the Russians. We have not had the time to investigate all the leads that come from such a complex case. We may find that more than just Russia is involved.

There was almost no public comment about the equally influential work by the Chinese. Their support for legislators in the U.S. with money laundered through political campaigns goes back much further than the 2016 election. The Chinese like to follow the laws of the country they are in, and have bought their way to U.S. citizenship using a campaign funding mechanism, the HB-5 Visa, strengthened and modified by Bill Clinton when he was president. The process is perfectly legal and carries few risks for those who do it. All it takes is money to make the process work.

In 2016, the *Washington Post* reported that the governor of Virginia, Terry McAuliffe, was the subject of an FBI investigation stemming from money given by a Chinese member of the National People's Congress who was also a lawful permanent resident (LPR) of the United States. How a member of the National People's Congress can become an LPR and live in the U.S. legally is an interesting question, but the more interesting question is how does that allow that person to give money to political candidates? The Federal Election Campaign Act (FECA) prohibits any foreign national from contributing, donating or spending funds in connection with any federal, state, or local election in the United States, either directly or indirectly. A foreign national is defined as any representative of:

- Foreign governments;
- Foreign political parties;
- Foreign corporations;
- Foreign associations;
- Foreign partnerships;
- Individuals with foreign citizenship; and
- Immigrants who do not have a "green card."

That last item is the key to how a member of the People's Congress, Wang Wenliang (later removed from his position in September 2016 for his alleged involvement in a pay-for-vote corruption scheme in Liaoning), can give money to the governor of Virginia, and two million dollars to the Clinton Foundation.[15] The Election Campaign Act has an interesting exception: "An immigrant may make a contribution if he or she has a 'green card' indicating his or her lawful admittance for permanent residence in the United States." So an investigation would be required to ensure the money did not come from the foreign corporations that were owned by the Chinese billionaire, some of which traded with the State of Virginia.

Governor McAuliffe and an unnamed number of "prominent Democrats" knew all about the EB-5 Program.[16] During an investigation of Deputy Secretary of Homeland Security Alejandro Mayorkas, the inspector general had pointed out the involvement of McAuliffe in this way[17]:

> Gulf Coast Funds Management Regional Center: Mr. Mayorkas intervened in an administrative appeal related to the denial of a regional center's application to receive EB-5 funding to manufacture electric cars through investments in a company in which Terry McAuliffe was the board chairman. This intervention was unprecedented and, because of the political prominence of the individuals, as well as USCIS' traditional deference to its administrative appeals process, staff perceived it as politically motivated.
>
> Mr. Mayorkas' actions in these matters created a perception with the EB-5 program that certain individuals had special access and would receive special consideration. It also lowered the morale of those involved.

The governor knew about the EB-5 Program, but he did not know as much as President and Mrs. Bill Clinton, or President Bush, who started it. The program started in 1990 as a way to get foreign investors to bring money for projects to the United States, but only 700 people signed up to use it while the Bush administration ran it. In 1992, it was modified to allow key provisions to be expanded. Foreign nationals who invest at least a million dollars in a new commercial enterprise (NCE) or $500,000 in a targeted employment area (rural areas and areas of high unemployment) could get LPR status and, after two years, apply for U.S. citizenship. The Clinton administration allowed the combination of jobs created to be aggregated in regional centers; the number of regional centers quickly went to over 700, and a few of them advertised their programs on the Internet. Most of the projects were located in California, where the adjudications of requests were processed, New York, Texas and Florida.[18] Some of the projects in California and New York were far from rural areas or areas of high unemployment.

In 2015, the *Atlantic Monthly* did a report on the EB-5 Visa program that exposed some of the abuses, including a gigantic Hudson Yards program that made 1,200 Chinese businessmen green card holders.[19] Of the 10,000 visa applicants in the year before, 9,128 were Chinese. This scheme allows Chinese citizens to become Legal Permanent Residents or U.S. Citizens; they are then eligible to contribute legally to political campaigns. There was no secret about why they did it, and a few Chinese were strangely honest in speaking with the press.

The CCTV Business Channel quoted the general counsel of Dandong Port Group, owned by Wang Wenliang, on the reasons for pursuing interests with the governors of states. He said, "One of the things I learned from this trip [the 2012 Democratic National Convention in South Carolina]: States have a lot of power. If you really want to influence ... let's say your China policy... it is really worth it to have emphasis on the state level."[20] The business relationships have managed to put money from U.S. companies owned by Chinese entities and newly created U.S. citizens to work in local elections, where they can build long-term relationships. Those relationships influence China policy just the way local politics influences national elections. And they do it within U.S. law, a characteristic of China that Russia seems to lack.

Rich Chinese are overwhelming the EB-5 Visa program, and the quotas are filling up earlier each year. The Justice Department indicted two women for abuses of the visa program and are looking for more.[21]

Although both countries influence the political process, the two governments are using considerably different techniques applied to both of the major parties in the United States. The Russians seem to favor the Republicans and the Chinese seem to favor the Democrats, though that might be an illusion because neither of them favor one political party over another. They favor influence over those their intelligence services think are most likely to rise in the political system or who benefit Chinese business. Their methods are consistent with the way their intelligence services have behaved behind the scenes. The Russians use blunt force methods to achieve their objectives; the Chinese are subtler. Both are effective in their own way.

The revelations of Edward Snowden must have gotten the attention of the Chinese, the first country to offer him safe haven after his departure from the U.S. What the revelations show is how far the capabilities to collect information from computer networks have come. For several months a document was published in a newspaper or magazine that related to some capability. By and large, these are collection capabilities of the U.S. intelligence agencies. Because Snowden also published the "Top Secret Policy on Cyber," we also know how the U.S. government characterizes those collection efforts. Because that document was compromised, the U.S. finds it difficult to define its responses to cyber incidents directed at the U.S. without addressing them directly. The secretary of defense has outlined a much more direct strategy against China, North Korea, Russia and Iran, including "preemptive cyber attacks" that might be a direct response or might prepare an infrastructure for a future attack.[22]

Snowden revealed quite a few secrets, but he also revealed how good the U.S. intelligence agencies really were. The documents disclosed show that the NSA had a variety of collection methods that, according to Stewart Baker, a former NSA general counsel, allow the collection of "virtually everything on the Internet." Our worldwide articulation of privacy concerns quickly fades away when a crisis on the scale of the attacks of September 11, 2001, shows deficiencies in the way that data was being collected.[23]

While disclosing how the NSA did some of that work was a great propaganda victory for both China and Russia, the sophistication of what was disclosed must have given the Chinese reason to stop and think. If the U.S. is that good at collecting information, maybe it is pretty good at making use of that information in a retaliatory strike or cyberwar. The Chinese were already familiar with Stuxnet.

Stuxnet was a form of deterrence, but it wasn't developed for that purpose. Since it wasn't designed to be discovered, its deterrent value was negligible, once discovered. U.S. allies would find the inability to keep secrets about this type of covert program, and the use of sabotage to achieve the political objective, as reasons not to ever engage in this kind of project again. Secrecy is of absolute importance to protect the allied relationships and provide deniability for the people who actually did the work. Sabotage is a word we generally do not use to describe the actions of that computer worm, but that is how the Iranians would see the intentional manipulation of centrifuges used to refine uranium into weapons grade material for a bomb that neither Israel, nor a number of U.S. allies, wanted Iran to have.[24]

Somebody was using other methods the U.S. did not like, such as the killing of Iran's nuclear scientists, so a better alternative was needed. Cyber strikes are "clean"; i.e., they won't leave a lot of blood on the floor of the place where they are used. It may have helped get Iran to the negotiating table if they believed there might be more to come. China has a different way of demonstrating deterrent strategy than the U.S. because they describe, through proxies, what "more to come" might mean. Just as the U.S. may have used Israel to help in its attack on Iran, China has done more to spread out their attacks through third parties such as North Korea.[25] This makes it more difficult to attribute the attack to China but allows China to measure the effectiveness of an attack of that type.

North Korea does not do much without China knowing. We are faced with nuclear bomb testing and threats to wipe out the United States, a clearly unachievable goal. It is a small, isolated country that stamped out a name for itself by threatening the U.S. and its allies anywhere it could. China can sit back and watch our reaction to that kind of manufactured crisis, getting the benefit of being able to anticipate what we might do if they were attacked by anyone else. When North Korea attacked South Korea in 2013, there was ample example of what could happen if China were to be attacked by the U.S., and the difficulties in attributing that to China. That attack disabled services in three of South Korea's largest banks, its two major TV stations and one cable channel.[26] It was only the first of the "demonstrations" by North Korea.

In what has turned out to be one of the more bizarre of those, North Korea warned the U.S. not to distribute a farcical movie about a plot to kill the Great Leader. North Korea was not just asking that the movie not be distributed in their country; they didn't want it distributed in the U.S. either, and they made threats against people who might go to see it. North Korea threatened to do something to retaliate if they did, and attacks on Sony were the result of that. The attacks were destructive; i.e., they destroyed Sony's corporate computer systems.[27] At the same time, they plucked information written by Sony employees and published that information on the Internet. These were private, internal e-mails that damaged the reputations of top level employees of the company. That is a warning to us. If you decided to try to retaliate against us, this is what you can expect. From a deterrence perspective, that is effective, but there were warnings on both sides.

David Sanger at the *New York Times* says the U.S. was buried deep inside North Korea's computer systems and could tell that this particular attack came from them.[28] If true, it would have been better for that capability not to be disclosed in a newspaper. The FBI previously had said it had evidence that the attack came from North Korea and advised the president accordingly.[29] Even so, the attribution was questioned by many outside government and could have been the reason for making further disclosures. So by denial, the North Koreans were able to discover something more important that the U.S. got in the exchange. The U.S. offered a tepid response, shutting down North Korea's small Internet domain for ten hours and applying financial banking sanctions to North Korean businesses.[30]

Both of these were pointless exercises if China really was behind the hacking of Sony. The Obama administration must have believed the North Koreans were not the perpetrators. When David Sanger wrote about the various options, administration officials said the U.S. considered retaliating against China. That included an option to do the same thing to some Chinese businesses that North Korea did to Sony. This type of retribution has a way of getting out of hand unless it is proportionate to the attack received. Had the U.S. not believed that China was behind those North Korean attacks, it would not be likely to have suggested retaliation against them. Yes, the U.S. also attacked North Korea to make the same point, but it knew what country was really responsible.

The Chinese launch more and more attacks because we have no deterrent capability to prevent them from continuing to collect intelligence of all sorts from our businesses and government services. In September 2015 and March 2017, President Obama met with the chairman of the Communist Party and president of China, Xi Jinping. The usual course of diplomacy is to work out ahead of time what will be said at the conclusion of the meeting and leak that to the press to show that something substantial will be done. Instead, the Obama administration leaked two other ideas: first, that there would be sanctions placed on China, similar to those used in Russia to deter them from going further in Ukraine.[31] Second, that the U.S. considered a variety of responses to Chinese hacking of U.S. businesses, and considered skirting the Great Firewall, or hacking Chinese government officials and releasing their e-mail, similar to what North Korea did to Sony.[32] Several months before and after his visit, the U.S. had done nothing. It almost seemed like the status quo was acceptable to both parties, and that the U.S. wanted to talk tough while doing very little. The Trump administration launched a Tomahawk strike while President Xi was in the U.S. visiting and moved an aircraft carrier and attending fleet moving closer to North Korea. That kind of action speaks louder than any leak.

The Obama White House defended their lack of a response as a means to avoid cyberwar with China. In fact, they were already in a cyberwar with China and had been for some time. Both sides know it. What that White

House was afraid of was that the American public would discover it had no will to engage, retaliate, or win that war. China was using the intellectual property of U.S. businesses to build itself into an economic power that would take over the markets the U.S. currently dominates, while protecting its own. That objective seemed to matter little to either the federal government or U.S. senior business leaders.

During Xi's visit to the U.S. in September 2015, a vague agreement between the U.S. and China created an arrangement that each should not steal business secrets from the other, but that we would, presumably, continue to collect intelligence on almost anything else. That is not an agreement that does any more than recognize the status quo. It did force the Chinese to change their strategy of using PLA forces to do the hacking and moved that responsibility to the Ministry of State Security, which employs contract hackers, is more skilled than the PLA, and is better able to operate without getting caught.[33] Had we used strategies like we currently have for cyber when we were fighting the Russians in the Cold War, much of Europe would still be dominated by a still-active Soviet Union. It should be obvious that we are unwilling to fight this kind of war. Something holds us back.

In August 2015, when several of the leaders of the U.S. intelligence community were asked about deterrence, the best answer for where the U.S. policy came from the Director of National Intelligence. In testimony before the U.S.

Figure 3. Terminal high-altitude area defense similar to that deployed in South Korea (Department of Defense).

Congress he was asked why we don't have a better strategy for deterrence and he said, "It's a political decision." We could do more, we might surmise from this characterization, but for the lack of political will to do more. No more incentive could be provided than the theft of 28 million security clearance records from the Office of Personnel Management, including the records of many senior business executives and government officials. That was the context of the DNI's words, yet nothing has been publically done to deter future thefts of government data.

What has replaced our concern about the thefts of data by China is Russia. That is not a coincidence. It takes the spotlight off of China and puts the focused attention of intelligence and law enforcement on Russia. Somewhere in those 32 agreements signed between China and Russia is a shift in attacks that will ease the pressure on China, which was becoming considerable. The U.S. knew China was behind the Sony attacks and had linked the People's Liberation Army with thefts of business intelligence in the U.S. China knew that the U.S. was blaming them for a range of thefts from the designs of nuclear weapons to patented information and trade secrets from businesses. In 2017, one can barely hear a peep about this kind of activity.

If we get into a tit-for-tat round of stealing information and releasing it to the public, there are several groups that would not do well in that exchange, but most particularly business leaders and government officials. Sony was a demonstration of that capability and the damage it can do. Releasing business e-mails can be both personally and professionally damaging to those who say things about their business relationships that they believe will never be made public. Imagine what can be done with the security clearance records. There is a good lesson in the Sony case for all of them. Political figures and business leaders have found reasons to avoid letting information about their conduct or reasons for taking certain actions into public view. The Chinese have demonstrated their deterrent capability and learned to be more careful to avoid detection.

Unlike almost any other country, the Chinese know who in their country is hacking someone else, or turning on other people inside their country. They have a disciplined national technology infrastructure with 659 million users, 80 percent using mobile devices, that is organized to allow the monitoring of every Chinese citizen. The Chinese show a willingness to support the kind of monitoring infrastructure that would be required with money, policy and other resources that would be impossible in other countries.

Their new anti-terrorism legislation directs the cooperation of any business from any country operating in China. It requires businesses to hand over source code and encryption technologies that protect communications from interception. The Chinese are good at recognizing an innate adherence to law that is found elsewhere in the world. They try to operate by the laws of countries they are in and modify their behavior to achieve their objectives without violating those laws—if they can. They also understand the idea that "it's the law" is a statement that carries weight in the business and government circles that operate in China. They make laws to achieve their objectives.

Though not all businesses comply, even under great pressure, avoiding doing so is getting harder and harder. The practical means of avoiding U.S. export laws that prohibit sharing certain types of technical information with other countries causes them to set up companies that are "Chinese" to manage the turnover. With corporate networks modified to be available to Chinese security services, businesses must wonder how they can keep their trade secrets secure within their own business environment. They should comply because "it is the law." Their problem, of course, is that law isn't reciprocal, nor is it well understood. At least some businesses adamantly do not comply with it, but find it more difficult.

The U.S. Chamber of Commerce in China in an attempt to help its members made recommendations to the Central Government that created some interest in the U.S. Many in the U.S. and elsewhere were not aware that China had been doing some of the things noted in these recommendations. These amount to non-standard policy manipulation to regulate foreign institutions:

- Continue progress in providing 30-day notice and comment periods for all draft laws and regulations across the board, as specified in multiple commitments.
- End the use of "window guidance" and release public directives instead.
- Improve comprehensiveness in the online publishing of all court cases within seven working days of a ruling as required by 2016 regulations.
- Improve transparency by releasing formal findings and case histories of anti-monopoly related investigations.
- Clarify customs and tax regulations so that foreign companies can fully comply and make more informed investment decisions.
- Provide written explanations whenever administrative agencies deny or provide conditional approvals for license applications or other approval applications, and adhere to decision deadlines specified in laws and regulations.[34]

In most countries laws and regulations are posted so those who are supposed to adhere to them can comply. But not only do the Chinese have vague laws, they also have a category of regulation called *Window Guidance*, a term that sounds odd because it is. It applies to unposted policies and regulations that are announced as needed, but enforced immediately. *The Financial Times* reported an instance of this kind of guidance that impacted banks in China.[35] In December of 2016, the regulators called foreign banks together and announced new currency controls intended to prevent the flow of capital outside the country, lowering the value of China's currency. There was no warning that this was coming, and the implementation was to begin immediately, trapping some banks in the middle of transactions. It caused disruption and confusion but accomplished what the Chinese wanted, stabilizing the currency.

The recommendation on court findings and anti-monopoly investigation is not new. The case of Rio Tinto illustrates how after-the-fact policy enforcement can be so interesting to companies trying to do business in China. Rio Tinto was told it was collecting business intelligence information on government-owned businesses and that the information was a state secret. Holding that kind of information could lead to a long sentence in a Chinese jail. Creating doubt about what can legally be collected puts companies collecting normal business intelligence at potential peril of being prosecuted if it were to be discovered. That kind of policy puts foreign businesses at a disadvantage in competing in China, which is clearly the intent.

These kinds of laws and policy documents encourage cooperation by foreign companies without saying what that cooperation is about. As a simple example, Trend Micro, a company making security software, has been told it must sell its Chinese division to a company in China. In a press release in August 2015, the company describes it this way:

BEIJING—(BUSINESS WIRE)—AsiaInfo Technologies (China) Co., Ltd. and Trend Micro (China) Co., Ltd., a wholly owned subsidiary of Trend Micro International, today jointly announced that AsiaInfo intends to acquire all of Trend Micro China's business, including licensing of product and technology rights within the China market. The acquisition represents AsiaInfo's commitment to emerge as the industry leader in big data and cloud security to help make the digital exchange of information more secure, reliable and intelligent for Chinese customers. This agreement will solely encompass business within China.

"The acquisition enriches AsiaInfo Group's information security product architecture, strengthening our advantage in both customization and integration," said Zhang Fan, president of AsiaInfo's security department. "This allows AsiaInfo to further expand into other sectors, such as finance, education, manufacturing and healthcare. The agreement will not only provide a broader market for future growth, but also protect national security through the creation of independent control of information technology development."[36]

Trend Micro can make it sound like this exchange was a good idea, but the last sentence of the press release makes it clear that China is prepared to buy out participants in its information technology infrastructure to make sure outside companies don't have "independent control" of parts of it. That lack of control puts companies operating in China at risk.

We will never know how, or if, the Chinese steal intellectual property from businesses that have networks in China, but business people who live and work there believe they do. The amount and types of controls on the Internet almost guarantee that Chinese intelligence services can access anything they want. Businesses operating in China have told me that they have tried to avoid moving equipment to environments controlled by the Chinese, knowing it will be stolen in those places. But it is becoming harder to avoid government interest and more painful to operate in China. So, instead of helping our businesses in China protect their information, we have abdicated the protocols of proprietary and export controls by bowing to Chinese law. These laws are constructed to do legally what we should not allow to be done at all. It facilitates theft of trade secrets and shrouds it in the name of national security.

A new report by Citizen Lab at the University of Toronto gives us some insight into how deep the Chinese have gone in monitoring their own population, and also gives an indication that they may be monitoring a good deal more than their own people.[37] Why does a browser used on smartphones made in China want to transmit these kinds of things back to a host server: user search terms, hard drive serial number, GPS coordinates of the user, nearby wireless networks (including a unique identifier) and places they visited on the web?

> The Windows version of Baidu Browser also transmits a number of personally identifiable data points, including a user's search terms, hard drive serial number model and network MAC address, URL and title of all webpages visited, and CPU model number, without encryption or with easily [decipherable] encryption.[38]

These are things the Chines are collecting on users of Baidu, the Chinese equivalent of Google, but there is more to it than that. Data such as this is also collected by third-party applications made with the software development kits (SDKs) provided through Baidu. Millions of Android apps are pushed over third-party systems to tens of millions of users. Why does Baidu, or anyone in China for that matter, need to know my hard drive serial number and the wireless networks around my home? There are only a few uses for any of that information and none of them are good. When Citizen Lab asked the question, they got an equivalent of "no comment." When Citizen Lab asked Baidu if it was a government requirement for them to collect this data, they got the same answer.[39]

What the SDKs do is allow the spread of this kind of collection behavior to third-party developers who might not know what the Chinese company has been collecting, and who may not have noticed that the kit they are modifying is transmitting this kind of material to China. When the kits are scattered all over the world, it becomes difficult to even discover where the changes to an SDK occurred.

The latest rounds of data thefts are in health care and the cyber security industry itself. While the U.S. federal government pushes the health care sector onto the Internet, more of that data is being stolen. We once thought that was being used for identity theft, which it has been, but there is little evidence that thefts by the Chinese are used in the same way as they would be by a criminal gang. This is health care data stolen for its use in cyberwar. It is collected because it is available. The Chinese probably know more about us than we know about ourselves. We have privacy and data exchange limitations internally to protect individuals. When that data is stolen, we no longer have that privacy, and that data becomes an asset for our enemies. Many of us do not see that part of the theft as an act of war, but that is because acts of war are not as clear in the cyber world.

Four years ago, we had just started to discuss what cyber attacks did constitute an act of war. The Defense Department, in April 2015, using a speech by the secretary of defense, tried to define when the U.S. military would react to cyber attacks as part of a "national response."

The Defense approach calls for U.S. Cyber Command to respond when there was "something that threatens significant loss of life, destruction of property or lasting economic damage"—which it characterized as applying to about 2 percent of current attacks on businesses and government. The purpose was to establish a deterrence policy that "works by convincing a potential adversary that it will suffer unacceptable costs if it conducts an attack on the United States, and by decreasing the likelihood that a potential adversary attack will succeed."[40] Knowing that the Chinese, and others, steal information from us doesn't mean that we have done anything about it. In fact, we seem to be in harmony with them. We complain; they ignore us and deny everything; yet we do next to nothing about their criminal theft of business information that drives the U.S. innovation engine. Charging Chinese military officers does nothing to discourage China from advancing their economic espionage using the People's Liberation Army (PLA). The distinction here is that China, as part of its national strategy, steals from the U.S. and uses that information to build industries of its own that compete with those same industries. The more serious aspect is our practice of allowing them to do so without penalty. We have no deterrent capability and no desire to stop them. Our business and government leaders have a lot to do with that. Both seem willing to support the theft of their own intellectual property by China, when it is not in their long-term national interests.

Senator John McCain, chairman of the Armed Services Committee, believes the military posturing does not work because it is not backed up by a national strategy. He says, "Make no mistake, we are not winning the fight in cyberspace…. Our adversaries view our response to malicious cyber activity as timid and ineffectual. Put simply, the problem is a lack of deterrence. The administration has not demonstrated to our adversaries that the consequences of continued cyber attacks outweigh the benefits. Until this happens, the attacks will continue, and our national security interests will suffer."[41]

6

Economic War

When the first edition of this book was written, it was more difficult to convince anyone that China was stealing business information from U.S. and global companies. Since then we have come to realize that China is stealing more than any of us had guessed. Stealing business information allows China to anticipate what commercial developments, mergers, and business strategies will compete with them in the years ahead. They do not have to guess what General Motors or Intel will do in their next iteration of designs, because they already know. They know where the technology is going, how it will get there, and what the vendors will be charging for it. In bidding for international contracts for energy, construction, transportation and communications they win, in part, because they already know what their competitors will bid.

But "China, Inc." is a hugely successful economic enterprise that uses some of China's best advantages to manage business development. It has a relatively good base to start from, but that base is eroding. Chinese wages are going up and since 2001 have risen 12 percent a year. At the same time, China's currency has risen gradually, and it occasionally fluctuates more than the government would like. Even with both of those things occurring, China still offers a cheap alternative to the wages and currency costs of Western countries.[1] While presidential candidates in the U.S. discuss variations on what the minimum wage should be, the Chinese pay less than one-fourth the wages of most Western countries, before those campaign promises are even realized. But there is something else that appeals to businesses, unrelated to production costs: one and a half *billion* potential customers.

U.S. business leaders know costs are less in China, and they want access to the local markets. To get it, they have to team with Chinese businesses to bring those products to them. Those products are not exported to China; they are made and sold there. To the U.S. with 316 million people, 1,357 million seems like a good market. But if that market consists of the people who are manufacturing the goods, the trade deficit suffers because we are not exporting what the Chinese are buying. There is no reciprocity in that approach.

It is no secret that we have a large trade deficit with China every month, around $30 billion. Although China's leaders have said they want to cut their reliance on exports as a source of their economic growth, focusing on internal consumption, it continues to run huge surpluses. In 2014, China's global trade surplus in goods and services reached $382 billion with the trade balance with the U.S. being the most advantageous to them. But the real amounts have not changed much since then: "In 2014, the U.S. goods trade deficit with China increased by 7.5 percent year-on-year to $342.6 billion, a record. And in the first eight months of 2015, the U.S. trade deficit in goods with China totaled $237.3 billion, a 9.7 percent increase year-on-year, raising troubling questions for the bilateral relationship."[2]

Table 1 shows the 2016 figures from the U.S. Census Bureau on the size of the Chinese trade deficit. We know that China gets the biggest share of our dollars for goods and services, so these numbers should not be a surprise to anyone. U.S. government officials tend to discount the deficit because China buys debt, $1.258 trillion, with part of their surplus.[3]

Yet the overall Chinese position is a little less clear, since it comprises

All figures are in millions of U.S. dollars on a nominal basis, not seasonally adjusted. Details may not equal totals due to rounding. Table reflects only those months for which there was trade.

Month	Exports	Imports	Balance
January 2016	8,212.1	37,145.7	-28,933.5
February 2016	8,048.7	36,160.8	-28,112.1
March 2016	8,952.3	29,852.7	-20,900.5
April 2016	8,667.1	32,973.2	-24,306.1
May 2016	8,518.3	37,535.2	-29,016.9
June 2016	8,822.7	38,579.0	-29,756.3
July 2016	9,156.9	39,487.1	-30,330.2
August 2016	9,393.9	43,247.3	-33,853.4
September 2016	9,559.9	42,022.7	-32,462.8
October 2016	12,698.0	43,807.1	-31,109.1
November 2016	12,119.3	42,620.5	-30,501.2
December 2016	11,626.0	39,381.8	-27,755.8
TOTAL 2016	**115,775.1**	**462,813.0**	**-347,037.9**

Table 1. U.S. trade in goods with China, 2016 (U.S. Department of Commerce).

other things in addition to treasuries. The Congressional Research Service summarizes their holding this way:

> U.S. financial securities consist of a mix of securities issued by the U.S. government and private sector entities and include long-term (LT) U.S. Treasury securities…, LT U.S. government agency securities, LT corporate securities (some of which are asset-backed), equities (such as stocks), and short-term debt. LT securities are those with no stated maturity date (such as equities) or with an original term to maturity date of more than one year. Short-term debt includes U.S. Treasury securities, agency securities, and corporate securities with a maturity date of less than one year. The Department of the Treasury issues an annual survey of foreign portfolio holdings of U.S. securities by country and reports data for the previous year as of the end of June.[4]

Internally, China makes it more difficult for competitors there by retroactively calling what U.S. businesses collect for business intelligence "state secrets." The threat, and a vague definition of what constitutes a state secret, makes local officials and company employees wonder how far they should go. Their competitors are state-owned companies to whom the same logic does not apply. This is standard for Chinese policy, which is light on definition but heavy on selective enforcement.

The Chinese use other methods to discourage companies from independently competing with state-owned enterprises in their own country. They have slowly squeezed out a number of potential competitors to Chinese businesses, although they have failed with others. The main form of this unfair competition is the use of domestic laws that favor Chinese companies and force non–Chinese companies to participate in either teaming arrangements or partnerships with Chinese businesses. They exclude large areas of the infrastructure from any non–Chinese involvement. In 2016, the government clamped down on companies doing business in China and forced them into agreements that give Chinese control to businesses that were previously 51 percent owned by others. These included companies such as HP, Qualcomm, Cisco, and Microsoft, which now have "Chinese companies" in China performing their business in that country. China claims it is because of concerns over foreign control of business equipment, a dubious claim at best.[5]

There is another obvious reason for calling these businesses Chinese businesses besides allowing them the ability to comply with Chinese law and ignoring that this arrangement can violate U.S. laws, particularly the export laws of the United States. The businesses operating in China know they need licenses from the U.S. government to provide a wide range of business information to Chinese counterparts. Some, for instance most large manufacturers, have gone to great lengths to show their actions to limit the types of technology that is exported. For example, Intel publishes weekly changes to its list of computer processors that are exported, the licenses that cover hundreds of those exports, and maintains licensing requirements that are voluminous and burdensome. Yet Intel also operates manufacturing, testing and research facilities in China (Beijing, Chengdu, Dalian, and three major facilities in Shanghai), Vietnam, and a number of other countries. A *Wall Street Journal* article last year indicated that Intel was moving its manufacturing of non-volatile memory chips to a new $3.5 billion company site.[6] Non-volatile chips retain information when power is turned off to the computer. The Chinese benefit from jobs and technology developments in their country but they are not as willing to bring partners in on other types of technology.

There is a formalized code for how these exclusions of business influence over infrastructure and technology are supposed to be implemented. The "Catalogue for the Guidance of Foreign Investment Industries 2015," jointly released by the National Development and Reform Commission (NDRC) and the Ministry of Commerce (MOFCO) and effective in April 2016, divides investments in industries into three basic categories: encouraged, restricted and prohibited. It further specifies how investments may be made and with what entities; e.g., an industry may require a joint venture or partnership be formed and a member of the Communist Party or Chinese citizen installed as the senior partner. Some industries require more than 50 percent owner-ship by the Chinese entity. Other businesses may be wholly foreign-owned enterprises (WFOEs). These laws too are vague and leave room for interpretation, but U.S. companies are learning to navigate them by feeling around for the "right" solution. According to a Pillsbury Law report,

> These "WFOE-permitted" industries include accounting (where the chief partner must be a Chinese national); the construction and operation of rail transit such as city metro and light rail; operation of performance venues; design and manufacture of transportation equipment such as aircraft engines and engine parts and components; airborne equipment for civil aviation and yachts; manufacture of electric machinery and equipment such as power transmission and transformation equipment; and manufacture and R&D of automobile electronic devices, such as manufacture of embedded electronic integrated systems.[7]

The prohibited list is odd in what it includes as much as for what it excludes in investments. The prohibited industries include these:

- Production and development of genetically modified plants' seeds
- Fishing in the sea area within the government jurisdiction and in inland waters
- Exploring and mining of tungsten, molybdenum, tin, antimony, and fluorite
- Exploring, mining, and dressing of radioactive mineral products

- Exploring, mining, and dressing of rare earth metals
- Processing of green tea and special tea
- Processing of traditional Chinese medicines
- Manufacture of weapons and ammunition
- Production of enamel products, Xuan-paper (rice paper) and ingot-shaped tablets of Chinese ink
- Companies of air traffic control and postal services
- Social investigation, e.g., surveys, analysis of views of a population, and the like
- Development and application of human stem cells and gene diagnosis therapy technology
- Geodetic survey, marine charting, mapping aerial photography, administrative region mapping, relief mapping, navigational mapping, and electronic compilation of common maps
- Institutions of compulsory education and special education, like military, police, political and party schools
- News agencies
- Publishing, producing, master issuing, and importing of books, newspapers and periodicals
- Publishing, producing, master issuing and importing of audio and visual products and electronic publications
- Radio stations, TV stations, radio and TV transmission networks at various levels (transmission stations, relaying stations, radio and TV satellites, satellite up-linking stations, satellite receiving stations, microwave stations, monitoring stations, cable broadcasting and TV transmission networks)
- Publishing and playing of broadcast and TV programs
- Film making and issuing
- News website, network audiovisual service, online service location, Internet art management
- Construction and management of golf courses
- Gambling industry (including gambling turf)
- Eroticism

In those industries where investment is restricted or encouraged, the cost of doing business has to be balanced against government interference.[8]

Samn Sacks, a China analyst in the Eurasia Groups Asia Practice, wrote that technology companies had to allow "invasive audits, turn over source code, and provide encryption keys for surveillance, and build local data centers," and that "counterterrorism law and banking sector information technology (IT) regulation both remain in play despite reports to the contrary."[9] Combine that level of oversight with the loose definitions of "state secrets" and there is ample reason to believe that China is not competing fairly in the markets, especially where it is creating its own businesses. What those business relationships lack is reciprocity.

The Chinese would say they are just trying to protect their own businesses, but there is more to this than protection. Their approach is to collaborate in domestic business to undermine the business relationships of global businesses that are not based in China. U.S. businesses know what the Chinese have done to regulate them, and that they are not competing on a level playing field. What the U.S. should consider is that the companies that still have major manufacturing operations in China work under these laws. To comply, companies are transferring technology to China and allowing Chinese industries to form unequal partnerships. We should be concerned about the long-term consequences of allowing China to operate on its own set of rules, unique in the world, and having other countries follow these rules without regard to international law.

If companies with major Chinese operations are complying with the kinds of requirements that its laws demand, the Chinese are getting their trade secrets, proprietary information, and access to data they process in amounts that are staggering. What makes that difficult for the allied governments is that most of what these companies do already violates export laws of their respective countries. Most of these export laws were made in advance of the Internet and have not been gracefully adapted to it. The Chinese know that and take advantage of it.

Two things the Chinese encourage are joint ventures and leadership by Chinese citizens in certain types of technical businesses such as electronics. This leads to "foreign businesses" doing business the Chinese way, or eventually not doing business there. The Chinese are not in a hurry to force anyone to follow their rules of engagement, but they are not afraid of offending foreign companies who do not do things their way.

In November 2016, the Obama administration made it clear how they intended to deal with the Chinese buying into U.S. businesses using state-owned enterprises. The question being asked by CFIUS is this: When a state-owned enterprise buys something, is it the government of China that owns the purchased entity or does an independent company own it? We must conclude, from the findings that followed that CFIUS decision, that it decided the government of China owns what its companies buy.

The widely reported sales that were disrupted included the Unisplendour Corp, part of China's Tsinghua Unigroup, Ltd., which planned to buy 15 percent of Western Digital. A month earlier Western Digital had planned to buy SanDisk, which makes a number of portable storage devices. Philips NV announced it would not pursue a bid by Chinese companies for a business making light-emitting diodes because of an inquiry by CFIUS. And the long-running sale of a German company, Axitron, operating in the U.S. was also undone the same way. The shift from specific deals by state-owned enterprises to specific sectors of the electronics industry in the U.S. narrows the scope of the CFIUS focus, but recognizes the principle of state ownership as a factor in purchases made in the United States.

The Chinese have allowed no defense to companies doing business in China. They are doing it by changing laws, eliminating any negotiation of compliance standards. They are careful to avoid exposure to U.S. security and regulatory agencies. The central government denies any such thing is occurring, but their denials are far less credible now than they were five years ago.

U.S. companies operating there are not doing any favors for their boards of directors, nor for the national security of their own countries, by cooperating with unequal policies structured to give access to trade secrets and proprietary data. Having our government allow these kinds of unequal trade arrangements is damaging to our national security, yet little has been done to China for the way it behaves. In the absence of cooperation, those businesses do not get access to Chinese markets. We can wonder if the pilfering of trade secrets is worth that access, but some boards of directors must believe it is.

Chinese businesses operating in the U.S. don't have to enter into arrangements with U.S. companies binding them to minority partnerships. They don't have to turn over encryption technologies and source code for software. Their business leaders will not be arrested for compiling competitive business information about their competitors. Our intelligence community does not steal information from Chinese businesses and funnel that back into the parts of the economy that compete directly with the U.S. The U.S. has to stop pretending this is "just business" and realize it is more than that. It is economic warfare, a large part of information war the way it is practiced by China. Government and businesses, on both sides of the situation, tolerate and perpetuate it.

In early 2016, Lourenço Gonçalves, chairman, president and chief executive of the mining and natural resources company Cliffs Natural Resources, said China was the main reason there was steel manufacturing overproduction in the world economy: "You can't call yourself competitive if your competitiveness is based on cheating the international rules of trade. Trade without fairness is not trade, it's war."[10] Apparently the Commerce Department and the Obama administration agreed with him, because they imposed tariffs of nearly 300 percent on rolled steel. This is almost unheard of in trade deals with China, but direct talks with the U.S. were ruled out, almost as if their opinions on the matter did not make much difference. There is only one thing worse in political disputes than arguing in public, and that is being ignored. There is potential for a trade war, but not for a long time. It is something neither country really wants, but that did not stop the U.S. from applying new tariffs to steel, and in March 2017, opening investigations of how steel and aluminum are being dumped on U.S. markets.

The business consequences of losing this kind of fight can sometimes be severe. In May 2016, Carl Icahn made headlines when he sold off his Apple stock because he was concerned about the relationship between Apple and the communist government of China. Apple says part of those troubles started when China asked for source code to Apple products.[11] New restrictions on media outlets put Apple in the position of closing its iBooks and iTunes movie stores in China. That had to do with the way China has changed the operation of companies that distribute books and movies. Apple's stock dropped in its longest decline since 1998, and Tim Cook, the CEO of Apple, went to China shortly thereafter. Whether he is able to accomplish very much on his visit may depend on his willingness to give in to Chinese demands or determine the costs of not doing that. Carl Icahn may be right from the standpoint of an investment strategy, but he is focused on the wrong companies.

What we should be more concerned about is companies that remain in China and turn over their source code when the government asks them to. We might ask Microsoft, IBM, Intel, NVIDIA, or Qualcomm if they have turned over their source code—or if they were even asked to turn it over. Source code is the future of any company and is the most proprietary thing a software company has. It is the way they make money; turning it over to another government is inviting trouble for the company that does. What they have largely done so far is to make their Chinese operations Chinese, so a U.S. company is not following the laws that contradict Chinese laws. China uses that source code to make their own counterfeit versions of the same product, to modify it and use it to collect intelligence, or to analyze it and make a competing product. A Symantec report in March 2016 indicates they use stolen certificates to make their software look like it was legitimate.[12] Companies that willingly turn over code are sacrificing their future for short-term profits. And they are doing it with a country that could easily be at war with other countries in a few years.

It is not possible to deal with a Chinese strategy to tilt the board in their own favor without doing something about it. Apple may have done something for itself, and the trade sanctions against China may have helped domestic production in the U.S., yet the reaction seems to be something less than the response required to counter a national strategy by a country that fights unfairly. For the most part, the U.S. does not fight with the same level of commitment that we see on the Chinese side, and it is unwilling to use the same tactics against this adversary. Playing with international courts and the World Trade Organization are largely political exercises that demonstrate "doing something" rather than actually accomplishing very much.

The U.S. and other countries have previously filed complaints with the World Trade Organization over practices that violate provisions of member organizations, thinking that might have some long-term effect. China is a member. Dumping, the practice of putting products on the market with substantial subsidies that reduce their cost more than would occur in free market pricing, is the most common kind of complaint against China. The U.S. filed 28 of the 48 cases against China in 2016. Most were anti-dumping cases.[13] In addition to those cases, the U.S. brought two major complaints involving the aircraft and automobile industries, plus an important one that nearly involved the entire computer industry. These are not the only places where China has acted like a poor partner to businesses, but they are current examples.

In a complaint to the World Trade Organization, the U.S. said, "China exempts the sale of certain domestically produced aircraft, including general aviation, regional, and agricultural aircraft, from the value-added tax (VAT), while imported aircraft continue to be subject to the VAT." This gives Chinese aircraft a competitive advantage in pricing in their own country. Aircraft manufacturing has a been a sore spot between the U.S. and China for a number of years, because China forced the U.S. aircraft industry to cooperate in the building of the Chinese domestic airline industry, something they did through joint ventures. This repeated the strategy used in the electronics and solar energy sectors. China rolled out the first commercial aircraft to compete with Boeing and Airbus in November 2015. The inevitable result of cooperation with China is the building of an industry that competes directly with those cooperating with them. We have to ask ourselves how a board doing good business practice can allow helping a country develop a competing product. It can't be "just business" when the outcome is an undermining of long-term profitability in the larger world markets. The idea of profits in the near term, exchanged for competition in the longer term, does not seem to make business sense.

In the last two years a number of aircraft component industries have joined with Chinese companies to build the parts of an airplane. Airbus and Boeing, the two largest aircraft manufacturers in the world, both have joint ventures with China, but more than that, some of their best known suppliers do too. Boeing is building its first offshore plane factory in China, building airframes in the U.S. and outfitting the interiors in China.[14] The Airbus A320 final assembly line, the first assembly line outside of Europe, is in Tianjin; it began operations during September 2008 as a joint venture between Airbus and a Chinese consortium of the Tianjin Free Trade Zone (TJFTZ) and China Aviation. The engines for the Chinese C919 commercial aircraft are made by General Electric in a deal that began in 2011. The arrangement included "sophisticated airplane electronics, including some of the same technology used in Boeing's new state-of-the-art 787 Dreamliner."[15]

The essence of the argument questioning this kind of trade is summed up by John Bussey in the *Wall Street Journal*:

> "It's unclear whether anyone in the U.S. government took a look at the GE deal in terms of U.S. competitiveness—the future of the aviation industry 10 or 20 years out," says an executive who advises companies working in China. He worries that a heavily subsidized Chinese jet program, enhanced with U.S. avionics, could eventually clobber Boeing. China has an incredible ability to distort markets, and we can't be reacting after the distortion has taken place.[16]

Clyde Prestowitz, a former U.S. trade negotiator who writes on global economics and business, says China is violating World Trade Organization rules that prohibit making technology transfer a condition of market access. "In a normal market the avionics would be done for that plane in the U.S. and we'd sell it to China," he argues. It is exactly that issue that makes trade agreements with China one-sided.

GE says it wasn't forced to give up its technology for market access. Instead, it sees this joint venture as a valuable piece of an existing global network of joint ventures and supplier relationships between the world's big aviation companies.

"Technology is the heart and soul of our company," says Rick Kennedy, a GE spokesman. "Why would we give away our future?"[17] That is a good question, and one that would have been asked of the board of directors and shareholders. Apparently, if they were presented with the idea, they agreed with Kennedy's position that GE was not giving up its technology for short-term profits.

Eaton Corporation, aside for being blamed by Hillary Clinton for moving jobs out of the country, has established a joint venture with China to design, develop, and manufacture fuel and hydraulic conveyance systems in the C919 aircraft.[18] Rockwell Collins and China Electronics Technology Avionics Co. (CETCA) have signed an agreement to establish a joint venture (to develop and manufacture the communication and navigation systems for the Commercial Aircraft Corporation of China Ltd. (COMAC) in the same airplane.[19] A joint venture between Safran S.A. and G.E. (Nexcelle) makes the nacelles for the engines; a Goodrich joint venture with China's XAIC will make the landing gear.[20] Parker Aerospace, in a joint venture with the Aviation Industry of China, will do maintenance, repair and overhaul of the aircraft.[21] Stanley Chao, writing for *Aviation Week*, rightfully asks if China can do well at building a commercial aircraft when it outsources almost all of its components. He also points out that the engineers at Commercial Aircraft Corporation of China come from building jet fighters, an entirely different type of aircraft than those for the commercial market:

> What is more disturbing is that COMAC engineers and managers are not interested in understanding the engineering discipline to build an aircraft. Rather, they only want the solutions to problems they have encountered. Because of pressure from the top, they will outsource the answers and solutions from Western companies and simply implement them without understanding the "whys." Meeting deadlines has taken precedence over good, solid engineering work.[22]

That may be true in the next five years, but in ten years China will have its own capability and understanding of the engineering to do considerably better—if we allow them to continue to buy the expertise to build their own domestic industries, rather than buy the products of U.S. engineering.

The automobile sector has an equally checkered but considerably longer history in China. In the last ten years the U.S. has filed several complaints with the WTO over Chinese practices in both auto parts and auto sales, with neither side winning much of any concessions from the other. In 2009, the U.S. imposed anti-dumping tariffs of 35 percent on tires being exported to the U.S. In 2011, China imposed duties ranging from 2 percent to 21.5 percent on imports of large American-made cars and sport utility vehicles, in part because bailouts of General Motors and Chrysler made their vehicles more competitive in the marketplace.[23] In spite of these government actions, neither side seems anxious to damage industry cooperation because it is to the benefit of both.

The most successful was General Motors' fifty-fifty partnership with Shanghai Automotive Industry Corporation (SAIC), known as SAIC-GM. The company was formed in 1997, and GM sold hundreds of thousands of cars in China after that. China accounts for 37 percent of GM's global vehicle sales, 36 percent of Volkswagen's, and 17 percent of Ford's, according to corporate filings.[24]

But General Motors is selling the first Chinese cars in the U.S. in 2016, somewhat of a surprise to the United Auto Workers.[25] China will initially produce the Buick Envision, a small sport utility vehicle, and a plug-in electric, the Cadillac CT6, which is made in the U.S. as a gas-powered model. Buick sold a million cars in China in 2015, so their near-term profits are offset by the sales of Chinese cars in the U.S. Volvo has already started making some of its cars in China, through a company owned by Geely Auto of China. Some U.S. buyers have professed surprise in finding their new Volvo was not made in Sweden.

The auto industry is somewhat different than the aircraft industry because China did not have a manufacturing capability for aircraft as they did for cars. The Chinese have been making aircraft for quite a long time, but their production facilities are relatively new to this manufacturing sector. In 1953, China's central government established First Auto Works (FAW) to build trucks. Since 2009, the automobile industry produces more cars and trucks than the U.S.[26] What the auto industry demonstrates is China's ability to produce a competing product on a scale that drives many others from the market. Whether that cooperation is good for businesses, expect China's, is another question. Nowhere is that more apparent than in the personal computer industry.

China makes the vast majority of all personal computers, including those of HP, Dell and Lenovo, the three largest sellers. But it also makes three out of the five top cell phones, including some from Apple and Samsung. More than that, it is difficult to find a server, hard drive or home router that is made anywhere else. As an example, look at the router on any Internet service provider such as Verizon, Sprint or AT&T. It will almost certainly say "Made in China." China makes everything in the network, and it dominates the markets in those pieces of critical equipment. The fact that China controls the manufacturing of these types of devices is a national security problem for any country that buys them.

Economics is not normally thought of as a part of war, but it has always been. In the big picture, if an intelligence service collects information about the strategy a trade delegation will use, and it provides that information to its own country's trade delegation, that is economic warfare. So, if one of the countries interested in hosting the World Cup decides it would like to know how all the other bidders are going to structure their bids, they can put the intelligence service on to trying to find out. A government can also collect intelligence and give it to its private industry or its Olympic bidding team. They could just as easily use government officials to try to influence the award of a contract for airplanes, a new national wireless system, or retrofitting of ships. Or a country can just lower or raise the value of its currency a little. The Chinese are better at this than anyone, although a few other countries might want to debate that.

There are grey lines here. One country might think using government officials to influence contracts is OK if they don't pay bribes to anyone. Another country night think bribes are part of business, so it would be foolish not to pay them.

This type of war with China started longer ago than you might think. At the time we were putting together the strategy for an information war, China was not ready to fight and the Russians were. The Russians will fight most anytime a war comes up, and they don't have a great track record because of it. The Chinese fight when the time is right for them. They waited.

We fought a Cold War with Russia and, I must admit, it was great. The military kept busy and the Department of Defense did very well during that time. Most of my career we battled them in one way or another and both sides were better for it. A defector told me once I was naïve about how the Cold War actually worked, and after he explained it, I understood that I may have been. He had been in Russia while I was over here in the U.S. We were fighting each other before he defected and started helping us.

He was in countermeasures, a dark business of watching what the enemy is doing and trying to come up with ways of defeating them by undermining their capabilities. If I fly into Syria's airspace, and I know what frequency their radars work on, I can build a jammer that will let me stop them from using those radars to detect me, until I can attack them. The Russians are pretty smart about these things so they want to build radars that won't easily be jammed so they make them hop around on the radar spectrum so my jammer can't pin them down. When I see them making such a thing, I want to jam more frequencies those radars operate in. This is a simple explanation of something a lot more complicated. Every frequency we jam, we can't use anymore, so the number of available frequencies gets pretty small after 25 years of this.

He asked me a question. "Do you remember that you used to tell your companies not to test their equipment outside when the Russian satellites were overhead?" I did remember that. "Well, our Russian defense forces used to tell us the same thing, when your satellites came over. Did we stop testing outside because we were told not to?" I assumed they did. "Of course we did not!" he said, slapping me on the back with a good-natured whack. "We still tested our things when your satellites were watching. Then your people would see what we were doing and they would start working on countermeasures for our weapons. Then your people would test outside when our satellites were over your head and we got to see what you were doing. If either one of us had stopped, the Cold War would have been over for both of us." Cynical, I was thinking.

The Chinese have taken the same principles and applied them to business, to make that business part of their war. They have a law that says businesses do not get a license and start operating in China just because they want to. China has rules, but they want to trade, and we know they do quite a bit of that. But they don't want to trade a few computers for an earthmoving truck from Caterpillar. What they want to trade for is the ability to make those earthmoving trucks. They want us to teach them to make the rope that they will use to hang us, which is one better than the Russians in Khrushchev's time. We have a large number of businesses that are willing to help them do it.

In the Cold War we understood what Russia was trying to do, so we cut back on some of our trade, particularly in areas where there might be some military benefit to them. We don't seem to see China the same way.

In order to get in the Chinese market, the business has to give them something of value, a technology that will be shared with their Chinese counterpart. China says it only does this in about 20 percent of the cases, but there is no way to say for sure. The percentages are not as important as the use to which the information is put.

When GM was negotiating to bring the Chevy Volt, their hybrid, into the country, they wanted to sell the Volt and have subsidies given to those companies that shared technology. What the Chinese offered them was a government subsidy of $19,000 per car. That is a lot of money. But GM has been in China for a long time, and they know their way around.

GM had built a car called the Spark, and China's biggest automobile maker, Chery, started building a similar car called the QQ. They look quite a bit alike. Admittedly, the names are not even close, but GM claimed the exterior and interior of the QQ looked a lot like the Spark. GM filed a complaint with the Chinese government that Chery had exactly copied the design of the GM Spark. The commerce minister said they did not provide "certain," meaning exact, evidence that it had been copied, which is not always like evidence you will see the CSI folks come up with on television.

Given that experience, GM said no to the subsidies for the Volt. That effectively increases the cost of the car by that $19,000, and makes it more difficult for the Volt to compete in the market.

The high-speed trains, one of which crashed a few years ago, were adapted by Chinese companies after Kawasaki Heavy Industries, Ltd., Siemens AG, Alstom SA and Bombardier, Inc., helped them develop the technologies. China says it has adapted and improved these designs and used them to compete. That is certainly a good reason for keeping the press from covering the train wreck. Their trains are faster and more efficient. The people they are competing with say the designs were used in violation of agreements established by their companies. China says it is great to be able to compete in this market. There are hundreds of similar stories about China's trade. They use it to steal technology and compete against the people they got it from. They think the world owes them the right to use whatever they get. It may not be fair, but it works for them.

Apple, the company that makes my Mac, iPad and iPhone, uses a third of a company called Hon Hai Precision Industry, also known as Foxconn. Foxconn has 800,000 employees, more than the combined employment for Apple, Dell, Microsoft, HP, Intel and Sony. They made the national news the first year this book was published and Apple called a press conference to take a field trip to the site making iPads. They also make my Xbox, my Intel motherboard and quite a few other things. They are not competing with anyone on what they build. They are good at it. They have produced a hot product, with high quality, probably faster than most other companies of the world could do, but it comes at a price.

Foxconn is probably best known for its Apple connection and for the number of suicides its employees have managed to accomplish in their Chinese factories. The factory that makes my iPad used to be an unhappy place. Eighteen people jumped to their deaths from high buildings in the Foxconn complex, and twenty others were stopped from jumping by nets put up around the building. They have labor problems every day, but no lack of labor, much of it from outside China.

Foxconn is not the kind of place that makes a person think of spaceship offices of the Apple campus, but Apple keeps it as its manufacturing arm anyway. This year they put aside nearly a billion dollars for a manufacturing improvement project that might relocate some of that work. The simple matter is, Foxconn keeps up with the schedule and demands of production, even though it may cause their people stress beyond what we might want or tolerate. We could use the same logic on child labor or prostitution, but we don't. Let some country try to make tennis shoes using children in sweat shops where they have to work 12 hour days and there will be an uproar like they were killing those kids. But we have a business were the stress is killing people and it is an internal problem for them. There must be a difference there somewhere.

The stress situation finally produced inspections by the Fair Labor Association, an industry group that helps to set standards for the industries they are inspecting. We saw everyone in a uniform on the Foxconn floor, and those uniforms all looked new. Maybe, in such a controlled society, it is the only way to get the wages increased and the working conditions improved, which it did. Apple is paying for that.

For a software example, the Chinese are learning how to make software from Microsoft. When the Chinese president visited the U.S. in January 2011, the current CEO of Microsoft participated in the visit. President Obama pointed out that 90 percent of Microsoft software used in China was pirated. This is probably not the way to start a state visit, but it does articulate the problem pretty clearly.

When Microsoft first started in China it was having a terrible time with pirated software and tried to deal with it as it had everywhere else in the world, by suing. They lost in court regularly, because Chinese law does not protect intellectual property the way laws in most other countries do. But suing gave them a bad reputation. Someone who sues to get their way, even though they weren't getting that either, is frowned upon equally in the universe. They got frustrated and went through five changes of Chinese leadership in their operations in five years.

The Chinese were starting to try out some new software, Linux, that was public domain. Linux is pretty good software and it has all kinds of applications that do well at most office functions. If the Chinese decided to adopt that kind of software, they might not have all the features of a Microsoft product, but they would have the right price. They can make decisions like that because they are a government-managed, centrally planned, one-party system. The U.S. could not even do this when it had a Democratic president and the House and Senate are both run by the same party.

So Microsoft decided having their software copied was better than not having their software sold there. This is a strange bit of logic. But what they also did was start cooperating with the Chinese government to help them build a software industry. Eventually, that cooperation resulted in government requiring the use of licensed software for itself, although the pirated software still is a problem for them. The other problem, of course, is that software industry they helped to create may produce software that competes with Microsoft. The decision to do that will come back to haunt them one day. We are seeing a similar turn of events with wind generators and solar panels.

Solyndra is already remembered long after the Obama White House finished defending itself against claims of acting in haste to give them a $535 million loan that now has to be paid back by the taxpayers. The Obama administration filed a complaint against China with the World Trade Organization because China's Special Fund for Wind Power Manufacturing required recipients of aid to use Chinese-made parts and amounted to a subsidy. After the complaint was brought, the Chinese stopped funding the subsidy. In the meantime, though, they were giving $30 billion in loans to their wind and solar companies, 20 times what the U.S. gives. This is the benefit of a centralized, managed economy. It is easier to move money around and control a market. If that fails, they are not above cheating.

American Superconductor Corporation (AMSC), which makes management software for wind turbines, filed suit in China in 2011, saying a government-controlled Chinese company, Sinovel, stole its designs for power management and competed with them for wind generation equipment. Following their latest update in 2013, China's highest court looked at jurisdiction, since the case was thrown out by a lower court. Sinovel made the turbines; AMSC made the power management systems that helped them work together. The person pleading guilty to giving the information to Sinovel was a Serb engineer working in Austria for the subsidiary of a U.S. company. This was an international case. He stopped working for AMSC in March but maintained computer accounts that gave him access. In April, Sinovel stopped accepting shipments from AMSC, claiming it was reducing inventory, and stopped paying for any more products from them. AMSC was clearly not happy about it. They said that senior employees of Sinovel actually paid for the goods that were stolen from them. Now, Sinovel has it all and doesn't need to buy it from anyone. With a few "improvements" like the high-speed trains made, they can be off and running. It doesn't bother them that the technology was stolen.

The civil suit filed in China listed the value of the goods lost by AMSC at $700 million, and share losses of 80 percent. Sinovel paid $1.5 million to the thief who took the technology. Sinovel countersued in China for $58 million for "breach of contract." This held the interest of the administration, which was said to have published a background paper on this case for the visit because it was "so egregious." In 2017, six years after the company began to pursue a solution in court, that case finally reached a higher court finding—it was rejected for lack of evidence.

While the Chinese learn to build airplanes the way Boeing and Airbus do, they have just started competing with them. Canada's Bombardier and Brazil's Embraer are the main manufacturers affected. These two companies must feel good about having Boeing and Airbus help China build up their competition while selling them the big jets. GE supplies the avionics for these aircraft, so they will learn to build them from one of the best. But the government intent is to compete with them one day, and judging from the past, that day will not be far off. We will eventually have an aircraft industry, automobile industries, drug manufacturing, software, and a host of others that will compete, because that is the only way China will have it.

To be fair, companies go into this with their eyes open. They know the rules, including those that say that a Chinese company must be selected as a partner to operate in China, and they know what they have to do to get into the markets there. They also, unless they get really bad advice, know their technology will be stolen. It goes with the territory. The Chinese have a different understanding of intellectual property. They think they should be allowed to use it, if they have it, no matter how they came by it. They don't feel too bad about opening up a complete counterfeit store, Apple logo and all.

You may think the Chinese are just good players of the game, so they win. That would be wrong. The Chinese make new rules to make the game harder for people who are playing against them, and they don't play by anybody else's rules. They cheat.

The U.S. has never had a policy to share its intelligence with the commercial businesses of the country, when so many other countries do it. With the businesses in China being state-owned, the distinctions are harder to manage. When a trade delegation complains that the Chinese were negotiating from our end position, they know those end positions were compromised. The reasons are varied, but a businessman traveling in China noticed his handheld computer had been compromised with software that would "phone home" if connected to another network when he got back to his office. Dr. Joel Brenner, National Counterintelligence Executive, said when a business traveler goes to China, he should have a throw-away cell phone, which cuts down on the opportunities to get into other people's networks—if you actually throw it away.

One network security specialist said some of his Fortune 500 clients traveling to China had software planted on their computers and their networks in the U.S. routinely mapped by the Chinese. This is not new, and Russians did the same thing before them. They probably don't see this as anything they should not be allowed to do. They believe everyone does it.

In a few countries of the world, as any world traveler knows, you cannot leave your hotel room without someone taking a look at what you have on your computer. This has become so blatant that most places hardly even try to hide the fact that they have been there. Planting software that phones home is a relatively new offshoot of that, but not surprising. It was a natural evolution of spying.

However, as trade becomes both an offensive and defensive weapon to exert influence, what is done in the name of keeping trade going has exceeded what we usually expect in the business world. There are other countries who do the same types of things, but for sheer in-your-face stealing, you can't beat China. This year was the first time in a long time that a senior administration official, such as the treasury secretary, acknowledged in public that the Chinese were stealing us blind. Some people see this as a very competitive nature of Chinese business people, but there are other names that can be applied to it. They want to win, because that is part of the strategy of economic war.

Global Domination

I was briefing a business organization on some of the aspects of this book, and one of the participants asked me what I thought the Chinese goal might be in their cheating to win in global economic war. I said, "Global domination," and everyone laughed. But in this case, not just those individuals but all of the business community needs to look to the objective. China wants to dominate in certain economic areas and has already said so.

What drives this latest emphasis is something called Made in China 2025. In the past two years, China has put $110 billion into mergers and acquisitions of high tech businesses. Dr. Robert Atkinson, president of the Information and Innovation Foundation, brought the point home in his testimony to the House Committee on Foreign Affairs:

> The current and emerging challenge will be around advanced industries that the United States currently leads or holds strong global positions in, because those are the industries China is now targeting for dominance. I urge you to consider what a world would look like in 15 years where U.S. technology jobs in industries as diverse as aerospace, chemicals, computers, instruments, motor vehicles, medical equipment, pharmaceuticals, semiconductors, and software and Internet are significantly reduced due to Chinese policies focused on gaining global market share in those industries.[27]

To win the war, the Chinese have to dominate in ways we understand. That does not mean that they only trade with the U.S., because the U.S. is still first in many of these areas. China makes a profit of about $300 billion every year, and in the zero-sum game of trade, they take that from the rest of the world. Sometimes, the rest of the world is not happy about it but is happy to have the trade.

We have to suspend our understanding of the Communist system to believe that Chinese business is just like any other business in the world. They are not like us. We confuse Chinese businesses with our businesses and they are careful to keep up that illusion. They incorporate subsidiaries in other countries. They are big supporters of teaming arrangements and joint ventures. They establish boards with members of the Communist Party in senior positions. They write bylaws and hold board meetings that can be seen by everyone. They have their companies act like they are independent of government control.

But even their public companies are not open in any sense of the word. Try and find out anything substantive about the managers of any Chinese company, and you will know that China is not like the rest of the world. There is very little to see, except smoke. The Security and Exchange Commission opened inquiries on its third Chinese company for what is a very complicated scheme that avoids the normal oversight of their companies that would come from operating a public company in the U.S. Everyone saw how this oversight can cause companies like Facebook to get trampled in the market. The SEC watches the practice of "reverse mergers," where a Chinese firm merges with a shell company in the U.S. so it doesn't get the scrutiny given to companies forming an initial public offering—particularly, the accounting. Third parties start trading in the shell company, raising its ability to get financing. A *Wall Street Journal* article says several of these "companies have had trading in their shares suspended or seen their outside auditors resign over the past year."[28] This isn't good business; it is potentially a criminal case that proves the Chinese are more than willing to do business outside the norm expected of a publicly traded company. They have done worse things.

It pays to know what the other businesses in China are up to and to have the political connections to smooth over conflicts with government officials. Ask Rio Tinto and Walmart.

Rio Tinto is one of the largest mining companies in the world and is based in the U.K. It does business in aluminum, copper, diamonds, iron ore, and energy and has 77,000 employees, some in China. In March of 2010, four of its employees were sentenced to 7 to 14 years for accepting bribes and stealing commercial secrets.[29] That last part is the reason for concern, since it is a fine line between a state-owned secret and a commercial secret when the businesses they were selling to were state-owned.

Rio Tinto admits they took bribes, which is more common in some parts of the world than others. In China, gift giving seems to be an institution among government officials and business leaders, and it is sometimes hard to make the distinction between something meant to influence and something that is like a business lunch or Christmas gift in this country. In the Rio Tinto case they took money and were expected to act in a certain way as a result. There is not much grey in that.

But, in most of the globe, there is a difference between business secrets and state secrets. If a person knowingly pays for either one, that usually isn't bribery. That can be theft of intellectual property or espionage. State secrets are usually marked in some way that identifies them as "Restricted," "Top Secret" or some other type of thing that can tip off a person that they might be protected by the government. China doesn't always do that, and to make it worse, tends to use the term "state secret" to mean "whatever we say." This makes it harder for anyone to tell, and has caused foreign companies to start getting rid of some of their documents, just in case they fit into the new category.

Doing business becomes much more interesting if you can't collect information from government-owned businesses or the government itself without violating some law somewhere, and that is exactly what Rio Tinto employees were charged with. The court said they caused China to pay more for iron ore than they should have had to pay. That part may even be true, but in some places that is called "smart business," not a crime. It is probably the same when they overcharge us for something they made, but you don't see any of their people going to jail for it.

In the end, they decided to charge them with stealing commercial secrets, not state secrets. People who steal state secrets don't last very long, and can have spectacular trials, when they are public. The trial still took several months, but it was low-key. They made their point.

China did the same thing with Walmart by arresting people and fining them $400,000 for selling pork as organic when it wasn't. I hardly know what "organic" is anymore, and they were arresting people for selling something because it wasn't. After one gang member got a suspended death sentence for forcing people to buy water-injected pork, you might have thought pork was more important there than in some places.

While the approach was supposed to be related to a food safety issue in China, after some really nasty chemically treated pork was being sold in other stores, it could be any number of things that were really behind it. Walmart got fined $500,000 for charging too much for certain types of goods and not doing their part to keep down inflation. It is more likely the continued pressure on foreign firms that makes it more difficult for them to operate there. The Chinese are glad to take investment money from them, expand their operations until they learn to compete with them, then tighten down their profits and take over the business. This should sound familiar to anyone who watches television. This is the Mafia business model, the "Tony" Soprano *modus operandi* with the Communist government being the senior leadership.

The Mafia was into all kinds of activity that could seriously get them in trouble. They loaned money at low rates to people they liked, or at higher rates to those who weren't family. They helped the business expand and allowed other people who were also family to buy into their operations. They branched out into legitimate businesses to handle money and cover their operations. They sent their kids to the best schools to have them learn how to do this well. They were subsidizing the best schools with so many of them going there. They kept everything in the family. It is cozy, and very, very communistic. (The Mafia probably wouldn't like that analogy, so I add that it was not my intent to imply that the Mafia is, in any way, communist.) This can be summarized in the interesting case of the 88 Queensway Group, a big, supposedly private firm.

When the U.S.–China Economic and Security Review Commission did a check of investments in Angola (they were trying to figure out if these were profit making business deals of strategic government investments), they found a few companies operating from the same address in Hong Kong, but they had never been linked in the press or business circles. A few individuals were controlling some small companies from the same street address, 10/F Two Pacific Place, 88 Queensway, Hong Kong.[30]

One person who was not well known in financial circles was a director in 34 of the companies. Her husband was tied to two state-owned companies, one of which was very closely linked to Chinese intelligence and served as a cover for agents operating outside the country. Another officer's residence was listed at the same location as the Ministry of State Security, which is home to China's foreign intelligence collection. Of course, this was a coincidence.

Several of the key personnel of the Ministry of State Security have ties to China International Trust and Investment Company (CITIC), China National Petrochemical Corporation (Sinopec), and possibly China's intelligence services. We have to remember that a close look at most of the businesses of the world will show some relationships like this from former government officials. They had jobs before they moved on the board of directors or became vice president of marketing for that airplane. This is what qualifies them for the position. In this case, though, there were some differences from a "normal business."

The group also had gotten high-level access to the governments and national oil companies of the countries where it puts its money. In order to get oil or construction projects in Angola, a company has to go through the Export-Import Bank of China and, by terms of those contracts, has to be mostly Chinese. The oil construction contracts gave guarantees of oil deliveries as collateral. This is a cozy arrangement for China and makes them the envy of many oil companies.

Two Chinese financing companies provide most of the money for those projects through arrangements with the Angola government's ministry of finance. Those companies have separate agreements for some other functions of other government agencies operating businesses in Angola. This kind of contact wording would attract attention almost anywhere in the business world, and probably says as much about the Angola government as it does about Chinese business.

Soon after it started some of these companies, the group began entering into joint ventures using some of its interesting connections in the Congo, Venezuela, Angola, and the Russian diamond business. The 88 Queensway

Group has established over thirty different holding companies and subsidiaries to do its investing. In addition to Angola, it has operations in sub–Saharan Africa, Latin America, Southeast Asia and the United States. In the U.S. it was briefly on the radar screen in 2008 for purchases of the J.P. Morgan Chase Building in Manhattan, 49 percent of the former New York Times Building, and 49 percent of the Clock Tower, also in Manhattan.[31]

Nothing tells more about the synergy between government officials, their spouses, and business dealings that enrich them than the growing case of a British businessman, Neil Heywood, found dead in a less-than-impressive hotel near his best customer and mentor, Gu Kailai, wife of a high-ranking party official, Bo Xalai.[32] Nothing has stirred the politics and business relations of two countries more than the flap that came out of this mess. It was historic.

Bo Xalai is not just another party official; he was talked about as one of the leaders who would take a seat on the nine-member council that runs the country in somewhat the same way the U.S. cabinet runs ours. We have more people and less power, but it is close enough for governments. Remember that this is a communist country and the centralized control is much more rigid than in the bureaucracies of the world.

Bo Xalai may not get that seat now because it appears, without accusing anyone, that his wife may have had Mr. Heywood poisoned, then covered it up by having his body disposed of before an autopsy could be conducted. This is generally frowned on almost anywhere, but here it seemed close to being accepted until the whole thing was upset by the local chief of police going to the U.S. Consulate in Chengdu. You can imagine the chief of police in Chicago driving over to Rockville to the Chinese consulate and turning himself in to report a crime. The Chinese would be dumbfounded and might take a few days to figure out what to do. That is about what happened here. After the smoke settled, anonymous reports started to come in, and a few have been very accurate. Someone close to the action is talking.

The police chief was a political embarrassment to the U.S because he was in a position to know what actually happened; his police actually investigated the case. He was eventually persuaded to seek shelter with Chinese and leave the U.S. out of it. This is international politics at its finest and has nothing to do with war, but it shows the lengths diplomats will go to. Diplomats want peace at the expense of any local official, though surely we will hear the whole wonderful story one day. For diplomacy to succeed, it must be wrapped in a package of friendship, with smiles all around.

He must have felt that he was expendable, abandoned by Bo and Gu Xalai and looking at a case causing an international uproar, since the British wanted to know what happened to Mr. Heywood. They were reading the newspaper reports and diplomatic cables flying everywhere, and they wanted to get more from the official sources. You might sympathize with the consulate, having someone like this showing up on the doorstop, but that is why they are trained to represent us.

The police chief proved impossible to cover up. People started to poke around and reporters began calling their sources. The more that came out, the worse it got. This started to filter up to the highest reaches of government when Bo Xalai was removed as Choungqing Communist Party secretary, the main base of his power. This is a little like the president removing the director of the General Services Administration, except that she would have been in the running for vice president in the next election. Bo's wife's power comes from businesses she operates. She was then under arrest, which seriously influenced how far her businesses would go, and how much "management" she could do from jail, where those skills would be tested.

The best information about this case does not come from the Chinese government, as we might expect, but from a website called boxun.com, which is outside the Golden Shield[33] (Boxun.com now has an English section that makes it easy to read). The site is hosted in the United States and operated by a fellow named Watson Meng, from the hotbed of political reporting in Durham, North Carolina. That makes it harder to control and much less responsive to censorship. However, as with most people critical of China, the site is constantly under attack. That is real journalism.

Meng's site is dangerous territory for the informants, who must be known to factions in the government. The Chinese government is making every effort to distinguish the killing as a "criminal act" and not part of any dealings the government itself was involved in. Bo was said to have been involved in this act, and so was removed, to be prosecuted. We see the same type of political response to the prosecution of the former presidential candidate John Edwards, who was accused of misusing campaign funds to support a mistress and his child. One of his aides even tried to claim paternity for the child to keep it from the newspapers. Bo is criminalized and the taint does not extend back to the Central Committee, which was willing to accept him until he became involved in a crime that nobody is accusing him of committing. He eventually succumbed to the ever-popular "corruption" charges and went to jail.

The political intrigue is about all we see, but the business fallout will not take much longer. Gu Kailai's companies will have less chance for contracts and trade that depended on her husband's name. In the business community, people stop inviting these folks to power lunches and those little get-togethers at the club. They forget names of relatives, friends, children and pets. There was some discussion about removing Bo's son from Harvard, even though he was supposed to graduate in the same month. His sports car might be downgraded to something a little less expensive. Those special privileges are the first thing to go, and it looks like the bandwagon is rolling.

The way this case is unfolding shows how politics and business are related and how quickly one or the other can be undone, when the timing is right. The Chinese might say, "This happens everywhere in the world," and they would be right, but other examples seldom lead to murder. It also is a clear light on the spouses of Party officials who seem to mix business with politics every day. It is difficult to separate the two.

From such humble beginnings comes greatness, backed by financing from the national government. By staying private, companies like those belonging to Gu avoid the disclosures required of most public companies trying to operate in those parts of the world, and ours. The Chinese understand the relationships between government and business and they are open about how it works inside the country. They keep it all in the family.

The 27 countries of the European Union (EU) got the same warning we did, on the U.S. Congress' attempt to start taxing some of the goods we get from China. When the Chinese need influence, they can get it, because they don't just hold Treasury notes in the U.S.; they have about a quarter of their money in EU debt, and they have promised to buy more. They have gotten bonds from Greece, Ireland, Italy, Portugal and Spain at a time when analysts would say these are a bad bet. Either they are the world's worst investors or they have something else in mind.

Europe has the same kind of objections to state-supported operations competing with their private business, only they use the World Trade Organization's anti-dumping laws that trigger when they get trade that hurts one of their local industries. As an example, there is a series of laws for bathroom and paving tiles, as hard as that might be to believe. The EU imposes tariffs as high as 73 percent on these tiles because they are sold at a cost the EU thinks is illegally subsidized and interferes with an EU industry. They have 49 anti-dumping measures they impose on Chinese goods and the Chinese do not like it very much. We have some, but not many.

Product certifications are the most interesting from the standpoint of protection of intellectual property. The Chinese require an inspection of the plant where the goods are produced and a certification of the goods by the Chinese government. However, in some cases, knock-offs of products will show up on the streets of Beijing before they ever are formally accepted, and way before they get into production. That has to be some product certification process they have there. It is clever, though. They manage to have a product on the street before a potential competitor can get started. That is certainly good reason to question the cybersecurity laws due to begin official enforcement in June of 2017. Trusting Chinese government officials to keep trade secrets seems to be too much to ask. Both the use of product certifications and security reviews give access to proprietary code to the Chinese government.

Investments in telecommunications are required by China's admission into the World Trade Organization, but of the 1,600 investments approved by Chinese regulators, only 5 had foreign financing. The interest of investors is the growth of their industry, adding 1.25 million cellular subscribers every week. If that seems like a big number, that's because it is. Only equipment manufacturers are allowed to invest there. The EU has complained, but the Chinese are saying it is partly a national security issue, and that is plausible without being entirely accurate. It is a national security issue, if it is reciprocal and is a national security issue to everyone. There used to be a saying of the Russians, "What's mine is mine; what's yours is negotiable." There must be a similar Chinese saying since that is how they operate.

China is Africa's biggest trading partner. Sudan, which is not exactly a garden spot of investment opportunities in the last few years, sells most of its oil to China on the same types of arrangements they established in Angola. It buys guns with the profits. The Chinese seem to be able to put the war between the North and South of Sudan behind them and live with the government, such that it is. They make headway because they are willing to ignore what governments do with the money they give them, and focus more on the trade they get in return. Most of that is oil.

The copper mines of Zambia have benefited from a $2 billion investment from China, but the new president has been critical of its mismanagement of labor. Chinese companies have ignored local labor laws, discouraged unions and strikes and kept pay low for workers. If they weren't Chinese, they would be called colonials.

China is growing business with Brazil and Latin America for raw materials, but the business has not always been good. Brazil has started anti-dumping tariffs on Chinese-made synthetic fibers, which they say are being sold at less than production costs, and has clamped down on illegal imports.

"Illegal import" is an interesting phrase because it is a pseudonym for "fake." Illegal imports account for part of the world's trade in counterfeits, and Brazil is not the only place with this problem. Creating fakes is a slightly different thing from stealing someone's technology and making the goods to compete with them. This is stealing the name of the company and making the product look like an original. We had a fellow in one of our offices who brought back some disks, made in China, containing almost every kind of Microsoft software you could think of. The product was the right color; it had an instruction manual with it; it had the hologram on it that made it look like the real thing. There were two glaring differences: there was only one box and there were several types of software in it; and it had every virus known to man on one or another of the disks. The U.S. claims it loses a billion dollars a year to counterfeit goods made in China, but they are not the only ones who get clipped.

The Chinese make Kawasaki and Honda motorcycles and "clones" of these, BMW, and police motorcycles. A clone is surely not something made under a license, though sometimes it is hard to tell what is made under a license and what is not. Their motorcycles look a lot like the ones made in Japan and Germany, only they aren't.

Everyone knows how the French are about wine. They are very discriminating and refined. When we buy French wine, we assume it is pretty good. Now the Chinese have come up with an interesting way of making the very best French wine, with, of all things, French wine, and it is good wine. They get original bottles from restaurants and copy the labels. Then they buy a good wine and put it in the bottle with the better wine label. Most of us are not French or could not tell the difference.[34]

Sixty Minutes has done a couple of segments on Chinese counterfeiting, and they show the scale of what is going on, just from a retail standpoint. The Chinese counterfeited a Harry Potter book that was never written by J.K. Rowling; they just made it up and used her name. They have a 5-story complex that sells almost nothing that is not counterfeit, from golf clubs to blue jeans. It is not illegal to sell "small quantities" of counterfeit goods, so they do that over and over.[35] There is some enforcement for the press and documentation of seizures, but the rest is corrupt. They say it is a cost of doing business in China. The manufacturers of the original goods don't think so, and neither do most of us.

Where this gets dangerous is in the manufacture of airplane parts and military supplies. When Callaway Golf Company started getting those fake golf clubs in, they discovered that they looked real, but they were steel instead of titanium, an important difference. The shafts were breaking and people were returning them to the place they thought made them. Airplane parts had the same problem a few years ago because some of them were breaking at bad times. It isn't amusing when part of the tail section jams while the plane is taking off.

Singapore, Vietnam and South Korea have concerns about China too, and there are two things that make that worth considering. First, it is rare for China to get pushy with its neighbors, but it has with Japan over the East China Sea islands, which are a long way from the mainland. Second, China is moving work to Vietnam for its cheaper labor, and they are not exactly enemies. This kind of thing rarely happens in the Friends of China Club.

Electronics

If you don't find it hard to pick up something in your house that is not made in China, then you are not trying hard enough to pick things up, or you took off all those tags that say, "Do not remove." I keep this book on a Smart-Disk, run by a MacBook Pro, with a Seagate back-up drive, connected to my Verizon switch and router, all made in China. The lamp, stapler, CDs and telephone that I use on the desk with all of this are also made in China. It actually made me feel better to find out that my printer was made in Malaysia, though that may be because it is old. HP told a person who complained about printers made in China not working well that all the printers were made in China, Malaysia and Thailand. None were made in the U.S. They did add, "Sorry." Of course they may be sorry that all printers are not made in China, or that the Chinese printers did not work. It is hard to tell which.

It can come as no surprise to anyone that the U.S. and China do not have trade that is equal, but you are not really seeing the whole story if you just look at the numbers of exports and imports. U.S.–China trade is not even close; and in the world of trade, "close" is measured in millions so the numbers do not look quite so big. It is also fair to say that we are pigs in the world market, according to U.S. Census Bureau figures, which are dutifully updated every month. We spend $500 billion more than we take in, every year, and that number is going up. That seems like an impossible number. If we keep going at the current rate, we will spend as much on the debt as we spend on National Defense. That should scare you.

Normally, when a country runs big deficits, its currency loses value. Countries such as China who benefit lose their advantage when their goods become relatively more expensive. If China continually runs huge surpluses with the rest of world—no, it isn't just with us—then their currency should go up in value. That also makes their goods more expensive. Our currency value is not going down and theirs is not going up, defying the rules of economics, at least as I understand them.

The reason that happened, according to the learned economists of the world, is that China controls the value of its currency; they don't let the market do it. This is actually pretty smart, considering the bad things that can happen with markets. Prices go up and down and imports vary from year to year. That doesn't happen in China. They control their economy, and they keep control of the value of their currency. It may not be "natural" to economists, but it sure makes sense to me.

China's currency is consistently undervalued (estimates range from 20 percent to 40 percent, with the higher number being used most often), meaning we can spend less and still buy Chinese goods. That sounds like a good thing, for us, since we can save money. But it goes back to the idea that the trade deficit is a difference between what they buy and what we buy, and right now that difference is big. The Obama administration has tried reasoning with the Chinese to get them to let their currency rise a little, and they have done that, but it is a long way from 40 percent and not nearly what we think they need to do. They have had the chance to label China a "currency manipulator," which is not something I have ever heard anyone called before, but they decided diplomacy is better than name-calling. Presidential candidate Trump said he would label China a currency manipulator on his first day in office. He didn't.

So, is China a currency manipulator and what does it mean if they are? The short answer is no. But they have been in the past, and that is one reason for the success of the China, Inc., machine. It was excused by the world's financial leaders as "necessary." Until 2015, China pegged the currency to the dollar at a fixed rate of 8.3–1, what the experts at the Brookings Institute call a "reasonable rate for a developing country."[36] Anyway we look at that, it is currency manipulation, but reasonable for a developing country. In 2005, China moved off that rate and allowed its currency to float. The currency rose 40 percent since then—more than any other currency. In 2015, China allowed it to fall for a short time and spooked the markets around the world. In 2016, the yuan dropped 7 percent against the dollar and China started pegging the value to a basket of currencies—the dollar, euro, Japanese yen, and pound sterling, and, beginning last October, the renminbi.

Since China does not have a good way to spend all the money it takes in by the exchange of goods—and probably would have a hard time spending it all anyway—they convert it to something that is backed by dollars, Treasury bonds. They have about a trillion of them, and since they are inclined to use third parties to buy into them on their behalf, some estimates go as high as 2 trillion. When the numbers get this high, it is relative. They own a lot of our debt. Surprisingly, Japan actually owns more.

This is like the bank that issues my credit card. They own my debt. If they decide to limit my credit, I can try to find someone else to give me another credit card or live with the limit. They can raise my interest rate, establish minimum payments, and send me threatening letters when I don't make a minimum payment. They could take it away if things got out of hand. They have leverage from this.

We keep the cost of our debt down by keeping interest rates low. If the demand for debt is low, interest rates will be low. If it goes up, it will cost us more to keep any new debt we have. China buys enough of that debt that if they stopped, we could have trouble getting other people in the world to buy it all, and those costs would go way up. So, if the Trump administration calls China out on this currency manipulation charge, China can say, "OK, we will stop buying your debt now, and maybe cut trade some too." This would probably not be good for us, but it wouldn't do that much to hurt them, even though we are a big customer. President Obama just said it would be better if they didn't do that sort of thing, and the rest of the world was just as unhappy about it as we were.

This little high-stakes poker game costs every one of us money, and by some estimates caused the entire jobs crisis we have right now. That reasoning is that the trade deficit represents what China has done to take away jobs, and those jobs would be here if they hadn't done those things. That is a little like saying more people should buy American cars so there might be more employment at GM and Ford. I like my Smart car, which seems to have been made in France, so that will not help the trade deficit with China very much. I have to stop buying things China makes for that to work.

I had this choice to make the other day at Home Depot. An extension cord for an electric lawn mower is not something you buy every day. They have gone up in price since I bought the last one 3 years ago. One of them was made in China and cost $32; the other was made in Mexico and cost $82. The one from Mexico had lights on the tips that told a person there was electricity flowing in the cord, and looked thicker and sturdy. I picked up the one from Mexico, and started to walk, but ran into the sales person who was helping someone else. I asked him, "What is the difference between these two?" "What are you using it for?" he asked. "Cutting a small patch of grass with an electric mower." He then bent down and looked at the label on both. He said, "The cheaper one will handle the load the same, and since you don't use it all that often, it will work just as well." Sold.

A former CEO of a major tech company has said that foreign countries don't take those jobs; we give them to them. There is probably more truth to that than the idea that they steal jobs from us. Boards of directors say they can make more money building something in China than in the U.S., and tap into their markets. I would tax them to death for doing that if I could be king.

GE is using that reasoning in sending its Healthcare Global X-Ray Unit to

China. They announced a $2 billion program to help the Chinese learn the medical imaging business the GE way. In exchange, they get to tap into that market of potential customers. From a business standpoint, they are right. Their profits go up; their managers get more money. If you believe the line in the 1987 movie *Wall Street,* "Greed is good," you can also buy the line that what is good for them is also good for the rest of us. If it were just about greed, we could ignore it.

The fact is, the Chinese make good things that we want, and they make a lot of them. They may not be as careful about safety, health care, small business rules, worker's compensation, and a few other things like that, but the truth of it is, we like what they sell. If the two extension cords were closer in price, I know which one I would have bought. They weren't. Maybe the reason they weren't is a lot more complicated than the thickness of the cables and the lights on the end.

A few of Washington's finest have said that we would be well off, with low unemployment, if it weren't for China's manipulation of currency. When it got down to doing something about that, Congress started introducing some bills to get action. We were going to start adding tariffs to some of their goods. Believe it or not, we do this every now and again and have over the last hundred years. We just don't get the kind of reaction we got this time.

Seemingly within minutes of this announcement, *The China Daily* mentioned that China holds a substantial amount of U.S. debt. Of course, the business world already knows this. A Reuters report says they picked up their lobbying efforts with Congress and the administration to help kill the bill. China can manipulate us to do what they want. That is what war is for.

The comments about our debt are a threat. They appear to be just pointing out the simple fact of ownership, but it means more than just that, and they know it. It is a warning. For several days after, they lowered the price of their currency, a little each day. We knew they could manipulate exchange rates, but now we know they will do it for effect. It doesn't look like much to the casual observer, but there were millions of dollars being made, and lost, in every one of those days. Don't forget that trade deficit. The loss would be ours. A little nudge here and there, with a deficit in the billions adds up pretty fast.

The Chinese steal our technology, rack up sales back to us, counterfeit our goods, take our jobs and own a good deal of our debt. They leverage those things to manipulate our business and politics. To those of us in Washington, D.C., it sounds like a normal day at the office. Only it isn't a normal day, and you are not seeing the whole picture, if you just focus on the economics of relations between the U.S. and China. If you are just focusing on those things, you might miss what is really going on.

The Wall

The China's Golden Shield Project got off the ground in 1999.[37] It is part of a larger effort to build up the capabilities of their bureaucrats to keep an eye on almost everyone in China. It is run by the office that does population control. I have to breathe deep just to think that there is an office of population control there, but that is an internal matter to the Chinese.

There are 12 separate initiatives of the Golden Shield, and like all government projects, they are all running a little behind schedule. In 2000, the Chinese Communist Party Central Committee organized a meeting with 300 companies, from a dozen or so countries, to talk about building a surveillance network that would combine the national, regional and local police and security agencies to monitor every citizen of China. That is scary.

The Golden Shield is supposed to construct databases of criminal records, fugitives, stolen vehicles, driver's licenses, migration data (that would be human migration, not birds), and a database of every adult in China. It includes geographic information systems (GIS), which allow them to geolocate a building or computer system, closed circuit television, which would see the place or the people located there. With such a system, it is possible to keep pretty close tabs on just about everyone, but especially those who might not be happy with the government. China would say this is an internal matter of no concern to us. I don't think so. Actually, the Chinese would say anything they do is an internal matter to them, but we have to draw some lines somewhere.

The Chinese are quick to break down their activities into smaller elements, each one by itself, justifiable in some way. Then they say, for example, "Everyone monitors their populations with cameras," and the people in Bonn, London, and New York will nod in agreement. They forget that this is a system that is trying to build a database of much more than just camera images, or find fugitives and terrorists by looking at those images. The Chinese use this system to keep people in line.

Several cities, such as Washington, D.C., have networks of cameras that can monitor the streets and major public areas, but none of them are trying to make a database of every person in the land. They do not target particular groups. If it were just in their own country we would not have much to say about it, but they don't stop there. That is where it becomes a concern to us, and not just an internal Chinese matter.

Control of the Internet

The Golden Shield is not specifically directed at controlling the Internet, but the Chinese control the Internet more than most countries would be able to do, by controlling the companies that provide the services that make it up. The purpose of this is to control the behavior of individuals. We saw a little of that in the Chinese request for Google filtering. The Chinese wanted to limit anything that smacked of pornography, not just politically sensitive things. The net they were casting was too big for any company to keep up with, and one would have to wonder how Baidu could do it any better than Google. Nobody assumes they can.

A user in China cannot access just anything. Twitter, YouTube, Facebook, and the *Huffington Post* are off limits, and so is anything related to the Chinese dissident groups.[38] If we were to do this, the Democrats could say we were not allowed to visit Republican websites, or anything related to the reform of the government, in any shape or size. We could go to websites of the Catholic church sites that are approved, but those would be limited to the ones controlled by bishops named by our government. We could not visit social websites that were not ours, and we would have our own versions of Facebook and Twitter. We could use only one browser, and it would limit what search results come up. This would be to spare you the hardship of sorting through all those things that could get you in trouble, anyway, so you should find joy in it.

Baidu had a visit from Chinese propaganda chief Li Changchun and Liu Qi, secretary of the Beijing Municipal Party Committee, who wanted "to learn more about the company's business and to give 'important instructions.'"[39] Both officials are members of the Communist Party's Politburo, which is made up of the party's top 25 leaders. This would be like having two cabinet secretaries visit Google to give them some direction on what they need to do to expand their business. Google might listen attentively, give them some coffee and donuts and send them back to Washington, but Baidu probably would want to pay attention.

The Chinese control other information that can be troublesome, partly by controlling their press and feeding the world with their own versions of things. The Russians used to do this all the time. They rewrote history, on occasion, but modern Russia controls its press as well as China. When one of China's high-speed trains plowed into the back of another one, there were not very many internal reports of it that were not controlled by the Central Propaganda Department. You can read the interpretation of events in Wikipedia and their own press, but once a story is out in the rest of the world, the wire services pick it up and start interviewing people who travel and live in China. Just compare the Wikipedia results with one of the world's wire services and see for yourself.

The Wikipedia report on high-speed trains in China says they lowered the speed of trains after the accident, so fares could be kept low. Someone actually modified the Wikipedia report twice since it was originally posted, as more facts emerge that could embarrass anyone reading their version. The truth would have worked just as well, but it would have pointed to the accident and reminded more people of it. Some managers of the train system were relieved of their jobs, but they were not relieved because of the accident; they were removed because they "were stealing money from the train system." This amazing coincidence occurred for each of the individuals fired over the incident. There was no mention of the train accident connected to the firing of an official. These are lies that are not even important to most of us. I have trouble trusting people like that.

The New China Daily posted these September 2011 extracts from various restrictions placed on reporting of incidents. They are not giving out guidance to young journalism students when they say:

- Regarding news about the "man executed by firing squad found 'resurrected' nine years later," no re-reporting or reporting is allowed.
- All Hunan media outlets are not to hype up the serial murder case in which the killer ate four of his female victims. From the looks of this one, they must not have tabloid newspapers in China.
- From the Central Propaganda Department: Regarding the fatal incident on train K256 of the Shanghai Ministry of Railways in which a passenger died after an altercation with crew members, all media outlets are not to conduct independent reports but to wait for the standard copy from the Ministry.
- Regarding Zhang Shichao's tortured-to-death case, no reports are allowed for any media outlets. He had been "helping the police" during a 70-hour interview in his office and died afterwards. His family said he had been tortured.

China could easily find out if anyone was violating the restrictions on censorship. We could say that this is none of our business since they are a sovereign country, and that could be the end of it, but the next time someone points to the superiority of this kind of system, think about whether it is the kind of place you would want to live. It doesn't seem like the kind of neighborhood that would appeal to me. They deceive their own people.

A War of Information

China controls its Internet because that is where information is. Quite a bit of information war is directed against computer networks, solely because that is where the information and communications are. This war is about networks, both as vehicles for transport and as storage of information about almost anything. Don't confuse criminals and this type of hacking. They overlap, but criminals are not usually working for a government as much as for themselves. They may have a government customer, if they are stealing something a government wants and will pay for, but they are not exactly government sponsored. Criminals and government hackers use the same techniques, so when they are noticed on the Internet, it is hard to tell the difference. The difference with China is that nobody operates a criminal venture on their networks without them knowing about it. If it is allowed to exist, they know it is there. It is to their benefit to allow criminals to exist as long as they support the overall goals of war. The Chinese use the term "patriotic hackers" to describe those how those individuals meet their needs.

We have seen network attacks go up dramatically in the last 5 years, but they were going up pretty steadily long before that. We are getting better at seeing this type of thing, and that increases the apparent numbers. The difference is in how they are being directed.

The computer attacks are much more "customer oriented" these days because they look for individuals and not just large computer systems. They can be accurate, and narrow, in who they target and how. This type of sophistication is needed because computer penetrations are so successful. There is too much information available, and too many targets. A little more focus helps to reduce the amount of time it takes to get a target that is worth having, and disrupt a user capability or deny that person a chance to act. This is the personalization of war, though it may not be personal to any specific individual, just a person in that position. If I'm attacking the head of NATO, I might not really care who the person is. I just want the office and a way in. This would start with a little look around.

When the Chinese hacked accounts of the McCain and Obama elections teams early in the presidential race, they were apparently looking for position papers that would identify who is writing the kind of stuff the president will read and how they think about these papers. We should ask ourselves whether, if the past is any indication, the Chinese were hacking the accounts of the teams of backers of candidates in the Republican and Democratic parties. President Trump, in an interview on CBS' *Face the Nation* in May 2017, said he thought that it could have been China, or any number of other groups, that hacked the DNC. None of the security companies or government groups that investigated the case have mentioned that as a possibility. But it seems likely that even if the Russians stole the information and published it on Wikileaks, the Chinese would not have passed up the opportunity to enlighten their leadership with papers and discussion points that helped candidates. They wanted to know that information and have experience with this type of hacking for intelligence.

In the same way, the hacking of Emmanuel Macron in France, by what is supposed to be the Russians, was probably not the only hacking of accounts going on there. Any intelligence service of a country with computers was probably doing the same thing. It is their job.

China has set up front companies outside China to allow them to influence U.S. elections by contributing money to candidates, without identifying the source of those funds. They will ask for favors and exert influence as any other business would. They look like legitimate businesses, and they are trying to influence U.S. policy makers. They can use inside information to do that more efficiently. They establish relationships that can be leveraged for other purposes and lay out the names of those to be monitored further. You can bet they haven't stopped doing that. This is a long process, but it wastes less time than collecting everything from random targets and trying to sort it all out.

China makes its internal systems safer from the same type of hacking by laying out a strategy for electronic warfare that includes rolling their civil and military telecommunications together. This gives them better opportunities for offensive and defensive operations. They seem to be doing the same thing in consolidation of military-controlled businesses and commercial businesses, in general. There is really nothing wrong with mixing the two together, but they won't admit they do it. They would prefer we believe their businesses are separate and independent of the state, the army and their intelligence services.

Of the 10 largest exports from China, the big-ticket items are related to network components, computers, integrated circuits, and cell phones. The free world is starting to look at the obvious potential for the use of those types of equipment for intelligence collection. The British, Indian and Australian intelligence services were said to have told their governments that there were substantial risks from equipment supplied by Huawei. India places limits on what their equipment could be used for.[40] China has complained that India has banned Chinese telecommunications equipment, in violation of the World Trade Organization rules, which China, to hear them tell it, follows very closely.

In July 2010, a *Financial Times* article said Huawei was thinking about buying the network infrastructure piece of Motorola.[41] If you hadn't heard this, it is because it was headed off sooner than the purchase of 3Com. This time, Huawei said it was going to use a mitigation agreement that would keep the business side from Chinese influence.

Mitigation agreements are used to keep foreign companies from getting control of companies in the U.S. that are doing classified work. I used to oversee the government side of the first one of these agreements, at Magnavox in Fort Wayne, Indiana. Philips, a Dutch company, was buying them. Magnavox did some classified work in their company and, under rules then, Philips couldn't buy them. To work it out, they came up with a mitigation agreement that put limits on what types of business relationships Philips could have with Magnavox. It was very detailed and very awkward to administer for both of them, as being the first to do anything usually is.

Mitigation agreements rely on two things: that both companies follow the agreement, and that the buying company does not influence how business is done in the U.S. part of the company. That second part is harder for some to follow, especially if they are losing money or have a real board of directors that follows the rules. The U.S. board can stop members from the parent company from attending certain board meetings, and that can cause some hard feelings now and again.

It is partly a "trust me" kind of relationship with the government. It is impossible to oversee every action. Every visit by the management of the foreign firm has to be documented, and the business has to be separated so that board members do not get to discuss the U.S. company's business unless they are from the U.S. company. It is hard to enforce and the Chinese know it, but what it also shows is their learning curve is short when they get shut off, as they did in the 3Com situation.

Zhongxing Telecommunication Equipment Corporation (ZTE) and Huawei are involved with the Nigerian mobile telecoms market, mostly through cooperation with existing vendors. Within a few months of Nigerian telecoms being deregulated, they both had offices there. Yes, they also get oil from Nigeria, so that worked out nicely. In a BBC annual poll on how people in various countries feel about each other, Nigeria was China's best friend. I would like to see who was interviewed in that poll.

The combination of China's existing global networks, its communications suppliers, its front companies and army-operated businesses adds up to an arrangement that is considerably different from the way we do business in the U.S. These extensions put deep roots in the telecommunications systems of the world and give access to the military to use them. Huawei, as would most companies, denies it has any attachment to the military; this may be accurate, but it cannot prevent the military from using its assets to collect intelligence and do other things too. In August of 2010, eight members of the Senate sent a letter to several senior Obama administration officials questioning Huawei's equipment sale to Sprint Nextel, asking these officials to respond to their concerns. Congress does not just pick out, at random, any company to complain about, but this letter did not go far enough in identifying where Huawei is making its inroads into the U.S.

Yahoo! and Alibaba are business partners, in that Yahoo! owns 43 percent of the stock in Alibaba. Alibaba has tried several times to get them to sell it back, but Yahoo! is not giving it up. A Japanese-based company, Softbank, owns 33 percent. Most businesses consider 5 percent enough to have some control over how the company operates, and Yahoo and Softbank together own over 75 percent of it.

Baidu, the Chinese "Little Google," has announced it was thinking about buying Yahoo!. It is a totally transparent idea to get Alibaba back by buying its stock and, probably someday, Softbank's too. This early warning means the Chinese have learned a little bit about the Committee on Foreign Investments in the United States and how it reacts to unannounced sales. They are giving plenty of warning. Baidu would then own Yahoo! and its stake in Alibaba. All would be right with the world.

Now that Google is out of the picture in China, Baidu will be able to keep a tight hold on the search and web activity of its population. Alibaba was probably finding it hard to keep knowledge of any government operations going on in their companies from their two biggest stockholders. If there were any hanky-panky going on, the government would have wanted that cut short, and putting it under another China-based company would do that. It takes a while to do, but it will get done.

Besides upsetting the business world, this shows to what ends the Chinese will go to control their Internet businesses. They had to know this was going to upset some people, and they did it anyway. They don't operate their businesses the same way we do, and they don't have the same idea of what the Internet is for. They do not have an open Internet in China, and their businesses are not like ours.

The Plan

Their strategy is called out in a plan. The Chinese system, much like that of the Russians, requires that their government issue a 5-year plan every so often, only their plan is one they actually try to follow. Our 5-year plans usually go out the window about 2 months into them. We don't use them very often and don't follow them as much as we should. They are just "guide-lines" for most people.

The Chinese plan their growth. Robert Herbold, Microsoft's former chief operating officer, said when he traveled to China, each place he visited started their briefings with an explanation of what the 5-year plan said, and how it applied to the work they were doing.[42] They tended to focus on three things:

1. improving innovation in the country;
2. making significant improvements in the environmental footprint of China; and
3. continuing to create jobs to employ large numbers of people moving from rural to urban areas.[43]

This causes excitement to all, no doubt. To most of us, this sounds like the State of the Union speech—"I want to have prosperity for all, health care for all, and no tax increases"—except that the Chinese start meetings with the plan, and they put their money where they need to meet their goals. They are building new cities to grow technology, and modernizing the ones they already have. They make communism look good. It would be hard to imagine a time in my government career when we started off a meeting with the Pres-ident's Goals for the Year. Maybe they have something there. Herbold cer-tainly thought so.

Reorganization

In 1998, the military was told to start getting out of businesses, partly because of the corruption; a process supposed to be complete by 2001. In this was more smoke. Just remembering that part of the profits from these ven-tures gets fed back into the families of these companies would make a person wonder how that would go. It has been ugly.[44] In 1996 and 1997 two anti-corruption campaigns were directed at the military and the police. In one of the exposed cases, police and customs agents were so heavily involved in smuggling crude oil that they actually were affecting the ability of the state-controlled oil companies to make a profit. Up to a third of the country's oil was being smuggled. When some of these operations were shut down, tax revenues went up by 40 percent. Three thousand businesses were turned over to local authorities to manage, and almost four thousand were closed. Eventually, exceptions had to be made, and besides businesses in railways, civil airlines, and telecommunications, the Poly Group and China United Airlines were at the top of the exceptions list. The army still operates from 8,000 to 10,000 businesses. Along with these changes, they started to reorganize.

In 2017 the PLA announced several changes in their operational and structural makeup to put the army under centralized control under the Central Military Commission (CMC).

Some PLA-run businesses were combined with ones that weren't, producing more confusion that keeps anyone from really telling what parts of industries are owned by the PLA and what parts are not. To add to this confusion, some of the companies are not real companies in the way we normally think of a business.

The FBI, in 2006, said there were 2000 to 3000 front companies being operated in the U.S, some by the PLA, some by Chinese intelligence services. This number is disputed, but the Canadians estimate they have between 300 and 500 operating there.[45] I would really like to see that list of companies that the PLA owns, even if there are only a few, but the wisdom is to classify it. If we tell them we know who they are, it will not take them long to change their name or move, but giving an accurate number would be something they should be able to do so that we all understand the scope of the problem. Not knowing who they are does not help companies that may buy from them.

Front companies are usually set up for a purpose connected with the business the company is in, but they are not very profit oriented. So, if they wanted to get into trading in Alaska oil, they could set up a small company in Alaska and start getting the right people together to make deals. The company doesn't have to make money, directly. It doesn't have to have board meetings or any of those other time-consuming things that are really painful for the officers. They just have to have minutes of those meetings—and I can do those up in an hour or so, without bothering anyone. They would invite Alaskan oil businesses to China to discuss exploration and extraction. They can invite trade delegations to visit Alaskan businesses. The trade delegations don't even have to know much about oil, if they have technology related to Alaska, and can spell o-i-l. They can arrange to visit places where certain types of technology are being used and sold. They follow U.S. trade laws for using "U.S. companies" for certain types of government work. Once they are set up, it gets easier to look like a real business.

A defector from Eastern Europe once told me that if a company looked perfect during an audit, I should spend more time there. He had worked in a front company once and liked it. It was hectic, because they were always trying to be two people at once, the guy who does the hiring and firing of a division, and the guy who is stealing from other business types. It was hard work. To make sure they looked like a real company, they tried to follow every rule of business, especially the rules of a government where they operated.

In 2000, a Justice Department indictment told how this works to siphon off technology. The Chinese have been operating "dozens of companies" with the same purpose, and a few with *only* this purpose. Dozens is not even close to 2000, so I have put that number aside and just looked at how they do it. I'm more inclined to believe the dozens number and hope the other one is wrong.

Front companies have been around for a long time, but not on this scale. Even 100 of them is more than I want to think about, and I really would like to know who they are. We shouldn't have to guess. Maybe we don't mind trading with them, but it might be better to know that we are buying that baby crib from a Chinese army company. I would like to know, anyway. I really want to know if I am buying software made in China.

In the last few years, they have gotten a little smarter about posting these types of management structures on the Internet. It is harder to trace army involvement in businesses because they have started to use names different from the parent companies and dropped military ranks to hide any potential associations.

We have to see China as a country that is not the same as us, and there is more than just the cultural aspect to think about. They are a centrally managed communist country with a plan to control as much as they think they need to. They manage business the same way, by using government agencies to develop and operate companies. They pretend to be the same as the rest of the world in how they do that, but they aren't. Combining military, intelligence and business is not usually something any country wants to do, because it puts too much power in the hands of the government. It seems as dangerous as it is.

Images of War

China should use nuclear weapons against the United States if the American military intervenes in any conflict over Taiwan. If the Americans draw their missiles and position-guided ammunition on to the target zone on China's territory, I think we will have to respond with nuclear weapons.
　　　　　　　　　　　　　—Maj. Gen. Zhu Chenghu[46]

This is a jolting image. Yes, we can see the mushroom cloud in our mind, children vaporized on the playground, and it is an image that we would not like to see in real life. It is one thing to parade around a J-20 on a runway, and another to talk about a first strike on another country as a part of policy. The Chinese deny that it is official policy of anyone in the senior leadership of government. "We have such a time controlling our generals," they say, hard as that is to believe. Let's hope that now that they have nuclear-armed missiles on their submarines, they can keep a little tighter control of them.

What a few people do believe, though, is that the Chinese would not care if half of their population died in a war like the one described. We say that we would not want to have the kind of casualties that nuclear war would bring, but we still have thousands of nuclear weapons, just in case. If we used a few of them, China would still have twice as many people as we do, although our numbers would drop quite a bit too. It gets out of hand pretty fast.

There are two totally false assumptions there. First, that the new China would not care if they had casualties that were half their population. They have jobs and houses and cars, and they are growing faster than we are. They may have problems now and again, but not enough to send them to a nuclear war. Second, none of us might be left after a real nuclear war. There are so many rads of radiation floating around that it wouldn't be good for anyone in the world, but least of all to those that were in the fight. We can see what happened around just one nuclear power plant in Japan to get the idea.

In information war, none of that matters. Truth is not as important as perception. Neither side wants that mutual-suicide kind of war, but the Chinese use the images of war, just like they use images of bad things that can happen to businesses operating in China, to make us think they are willing. They know we believe they wouldn't mind the losses. They do this over and over, with consistent, repeated images, particularly with Taiwan. Donald Rumsfeld says in *Known and Unknown: A Memoir* that when he went to China for a visit, they took him to a public exposition to see a model of swirling fighters and ships ablaze—our ships in Taiwan, with the Chinese attacking them.[47] It wasn't just for him that this was done, but it was nice that he was there to see it.

The most clever of these images are things they don't have very many of, like stealth fighters, submarines, and aircraft carriers. They show us one of a new generation, just one.

7

Real War

At the same time, we have real concerns about war. China is building up their military in the old-fashioned way.[1] The army has 2.8 million men and women at a time when we are talking about cutting ours back to under a half a million. This isn't exactly a fair comparison, since they lump some functions in the military that we would not. They have a large national police force that isn't included in that number. Officially, their military is used for "local" combat operations, which can be anything they claim as their territory. That's why those lines they draw on maps are important to countries other than Japan and Taiwan. We don't like to see Japan included in the territory that China says it will defend.

Their concept for the use of military force is "active defense." Attack only when attacked, but operate offensively. This means that it is OK to throw wood in the water in front of a ship to see what happens and how far the enemy wants to go. If they attack you for it, it is OK to respond. It may be twisted logic, but they can always say they were attacked and responded to the attack.

They also believe that a response is not bound by time or space, so it does not have to happen right after someone turns on the water hose to get them away from the boat. It is all right to wait and to strike somewhere else where the enemy is less prepared or does not respond in a way that allows for any kind of force to be used. This doesn't work with training dogs or kids, but it may be different for countries.

The application of force is generally with the army, where increased emphasis and money has come in three places: nuclear offensive forces, space warfare and cyberwar. Their offensive nuclear forces consist mainly of missiles, much like most other militaries of the world, with ICBMs that can reach the entire U.S. Some of them are on ballistic missile submarines. They are trying to build missiles that are more effective against anti-missile systems like those the U.S. has. Now that they have all of our nuclear weapons designs, they can have nice missiles to put them on. They have threatened to do just that.[2]

They have a "no first use" policy for military nuclear weapons against other nuclear countries that is a little cloudy. Most countries, like us, just come out with it. "We will not use nuclear weapons first in a conflict." Some of China's military believe "first use" applies when the country is threatened, when nuclear force is threatened by the enemy, or when the other side's conventional forces look to be winning. They also do not see a nuclear weapon detonated in the atmosphere instead of on the ground as being the same as a first strike. All of this makes it more difficult to figure out what to do in the event they want to go to war.

We certainly don't qualify first use; we say we aren't going to use a nuclear weapon first. The conditions they lay out are things that are not easy to decide, but where national policy is important, we don't like fuzziness. That kind of unclear policy has an effect on everyone else. It kind of leaves their options open, and keeps everyone guessing. It is a deception.

The benefit they get is making other countries hesitant. Keep in mind that there are not too many countries in the world with nuclear weapons, so we are not talking about making the rest of the world nervous about it. Just a few countries will be paying attention.

War in Space

China had 15 space launches in 2010, a national record, and they have a program to get something on the moon, whether an explorer or humans. This is the first year that they equaled the U.S. in launches. They have developed anti-satellite weapons and may have intentions of using them against both communications and spy satellites. They practiced by using one on an old Chinese weather satellite, and the U.S. has accused them of using lasers to blind our satellites.[3] They have just launched their fifth GPS satellite, which can mean a number of things, but mostly that they want their own rather than having to use someone else's. They may just want to have some, if they decide to shoot all the others out of the sky. Branching out into space is relatively new for them, but it is part of the homeland, just higher up.

Our military certainly believes wars will be fought in space. There are commercial, military, and other government interests that go far up into the sky, and if we are going to protect them against other militaries, we need capability there. If they start taking down satellites, we are going to be awfully close to war, but there is a good question about what to do about it. We could respond in kind and knock down one of their satellites, and both of us could end up without any. Our TV and high-priority phone service is going to be limited if that happens, but is it really war? I don't know. We certainly have to think about it because they have shown us they can do it.

The Chinese used business partnerships with Loral and Hughes to obtain technology to improve their ability to launch satellites. Both acted to help China improve its missile launches, and they did it knowing they did not have the required licenses to transfer information. As ridiculous as it sounds, a bilateral agreement gave them permission to launch our satellites from launch facilities in China. Whose bright idea was that? Unfortunately for us, "technology controls at the launch sites" were not very good, and the Chinese "probably benefited from access to these satellites."

There is a difference between spying and what happened with Loral and Hughes. When I was still doing industrial security, we had a break-in at one of the facilities I inspected, and a person was found there with his hand in a container that had one of the company's prototype things in it. He was a spy, taking a big chance. Stealing from a company with fences and armed guards is really risky. But in the Loral and Hughes case, the Chinese had access to the object they want to see, and we allowed these companies to send the satellites over to them where they can look at them at their leisure. On top of that, Loral and Hughes were helping the Chinese improve their launch success by giving them information they were not allowed to have.

The House Committee report concluded that "U.S. policies relying on corporate self-policing to prevent technology loss have not worked." This is an understatement. I still have a hard time figuring out why our own government would want to turn our satellites over to another country for any reason, and just as hard a time figuring out why a company gives information to them that will only make missiles work better that could be targeted against the United States. Both of them were our defense contractors while they were doing it.

The Defense Department used to have a program to inspect defense industries and help them maintain some protection standards for information they get from the government. Twenty-two different agencies used that function to keep an eye on their contractors, but it was largely phased out. We need to bring it back and start looking at how this information is being protected when it is given to them. Having that information in computer systems that are in China makes that a very difficult problem to control.

Cyberwar

Cyberspace is the area that gets the biggest expansion in the Chinese build-up of defense. It combines elements of regular military and commercial telecommunications companies and uses different types of weapons.

Although information warfare has been around for a long time, cyberwar and hacker war are relatively new. Computers were not around much before World War II and did not really get networked around the world until the 1990s. So when a person claims to have 30 years of experience in information warfare, it is always a good idea to ask them what they were doing in the time before networks.

The first time many people in the U.S. became aware of this type of warfare is when they noticed the problems Google had in China. We first heard that they were being asked to filter their search results so that certain types of dissident groups would not show up in the search list. If I did a search of church groups, the Falun Gong would not show up on it. The list of things was pretty long, and Google objected, then eventually moved to Hong Kong, where they would not have to do that type of filtering.[4] At first, they redirected any mainland China search to Hong Kong, automatically, but they eventually backed off of that, giving their users a choice of service. The Chinese said, "If Google wants to operate here, they have to follow our laws." That was pretty fair by most standards and certainly not war-like in any way. That could have been the end of it, but it wasn't.

Most of the readers of the Google news did not know that the person behind Google in China was a former Microsoft employee, Dr. Kai-Fu Lee. He was born and raised in Taiwan. When Google hired him, Microsoft sued to prevent him from telling anything related to Microsoft, and there was some back and forth on this before it was finally settled. The good doctor did some of the planning and recruiting for Google, but they had problems operating in China from the beginning.[5]

Google China had unexplained outages in their website, while their chief competitor, Baidu, did not seem to have the same problem. Maybe being in a different city was part of that, but it's not likely. This is not "fair" in any type of business deals, but the Chinese have not been known for fairness in competition. Google's public relations manager was fired because she gave iPods to senior government officials and billed them to Google. It was acceptable in China to do this, but not in the U.S., so Google fired her and the person who approved the purchase of the iPods.

There was also a running gun battle with the Chinese on web filtering that seemed to be endless and trivial, and it got worse when the Chinese hosted the Olympic Games. They decided it was important to do more filtering and that they should filter the Google.com site and the Google.CN site. Google.com was in the U.S., and Google thought this was outrageous. The Chinese did a demonstration of some of the Google search results to show that pornographic material was being displayed by some of the search results.

They promised Google would be punished for this. Nobody could have imagined what they meant by that.

We see Google as the good guys, the guys who Do No Evil. So, in 2010, when China started hacking into Google accounts to try to get access to some information on dissidents, we felt like Google was being treated badly. Google didn't like it either. China crossed one of Google's red lines.

China started looking for people in the U.S. who supported human rights groups in China, and they led them into places we were not happy about. That was when they hacked my congressman's office. They broke into Google's e-mail accounts, called Gmail. Google really didn't like that at all. They said they were going to stop doing filtering of their websites. Everyone there knew what this meant. Dr. Lee decided it was time to leave Google, so he knew what it meant.

The Chinese must find this hard to understand, since the use of their Internet companies to control dissent is national policy. They believe that what they were doing was perfectly acceptable and should have been recognized as such by the rest of the world. They should have known we were not the same, but they let it get out of hand, when it didn't need to be. Even a little bit of stretching would have allowed us to accept the idea that they were OK in thinking that it was acceptable to use the Internet to control the population, if they only did it inside China. They didn't do that either.

Most people chalked this up to China looking for dissidents and getting a little out of control with it. This kind of thing happens in places like Myanmar and Tibet, so we should assume it is going to happen and move on. It could easily have ended there, and probably did for most of the people in the world. They stopped thinking about it.

Except … there is something odd that happens when this type of case comes up. The computer security community starts to look around for similar attacks, using the same type of techniques, and to look more closely at the amount of damage done in the original. This takes time to do and be accurate. Even a small investigation can take a week, and big ones can take several months. Human beings can forget something important in a day or so, so the hacking of Google was long out of their minds by the time this one was over.

It turns out the good guys were not just getting hacked by people looking for human rights advocates in the U.S. The techniques used to get into the accounts were common in a number of places, and the attack was against more than just human rights. They were stealing source code from companies like Adobe, Yahoo!, and Dow Chemical. What kind of source code, we will probably never know, because none of them are talking. It turned out 34 companies in all were involved. The more we look, the higher that number is. There is no way this was related to the problems Google was having with the government. It had been going on before any of that started.

Source code is the code that human beings write to lay out instructions for a computer or a network device. It is usually considered more important than the code that comes with the computer, because it is the original. All the other comes from it. Source code can be modified and it still looks like the same software running on any of those other computers, but it isn't the same. It may do other things in addition to what it does on my computer or it may not do things it is supposed to. The Chinese can certainly write their own software, so why would they even think about stealing it?

One reason is it takes less time to make your own software if you have source code from somebody else. It cuts the development time down from a year, in the case of a really complicated thing, to a couple of months—or less, depending on the skill of the people doing the work. Let's be clear here: this is illegal, but it does cut down development time. Thousands of lawsuits pockmark the legal landscape of software over just this type of thing.

A second reason is the ability to make software look like the real thing but having it do some things that it wasn't supposed to do. Hackers seem to do this all the time now, though they didn't use to think about it very much. At a simple level, we could modify our mail program to look for "Falun Gong" in an attachment it was sending to someone else, and if it found that name, send a copy to another location without tipping off the user. The possibilities are endless, once you have the source code. By the way, try to find something about the Falun Gong and you will see how successful the Chinese have been on limiting access to anything about them. They have done pretty well.

Now the security community sees that the attacks were not limited to the Falun Gong and other dissident groups. Now they want to try to find out where else these folks have been and what they have been looking for. They found more. The techniques were not exactly the same, but they used the same principle. Send a document that has embedded code in it, like a Trojan horse that can take control of a computer, and make it look like something that everyone wants. Hackers used to do this all the time with pictures of naked women or movie stars with clothes on. The subject matters have gotten more sophisticated over the years, but some of the others were more entertaining.

The Chinese were promising things like the new list of military base closures and copies of budgets that had not been released. Although people must have wondered why they were getting something like this without asking, most of them would open it anyway, believing they should not frown on good fortune. The embedded code executes and the computer is open for the attacker to use.

While the good guys were looking at how this was done, they found something called Ghostnet, a China-based network used for hacking. About all anyone can authoritatively say about this network was in two reports by the *Information Warfare Monitor*, published a year apart.[6]

In the first report, they said they were not so sure that China itself was involved and that the spike in Internet hacking from China could be due just to a 1,000 percent increase in Chinese users over the last 8 years. Maybe somebody else did it and the Chinese were being blamed. Maybe there was a natural explosion in hacking, given the increase in the number of users. All of these are possible. With those kinds of numbers, anything could be behind it.

They had penetrated enough systems that many of the ones discovered during the analysis done for the second report were compromised during the first attacks but had not been discovered. They had been deep into these systems for a long, long time. In addition to the cyber spying, the Chinese have been engaging in a broader spectrum of human collection. A *Wall Street Journal* article on the subject was a reminder that the computers are not the only good vehicle to get information of value and the Chinese continue the long tradition of stealing information with human spies.[7] There are three now on trial, or pending trial and two of them were recruiting contacts within the CIA. The most current is Kevin Mallory who worked for the CIA and, according to the indictment, gave classified information to the Chinese in exchange for money. The Chinese supplied him with an encryption system to transmit more data. He had seven documents on a secure digital card and called his wife to find it before someone came looking for it. Unfortunately, that call was made after it had already been discovered by the FBI. The Chinese follow a time-honored tradition of attempting to occasionally recruit people who are of Chinese heritage. Perhaps because of the belief in how the Chinese intelligence services recruit, the U.S. law enforcement community has had a tendency to focus on Chinese nationals and recent resident aliens, when those are not the only spies being recruited.[8] However, if one follows the Justice Department press releases on indictments, it does seem that a majority of persons accused of industrial espionage for China are of Chinese descent, and the FBI and other intelligence services are getting better at detecting them.

In October of 2011, General Hayden, who used to be director of the National Security Agency and was in charge at CIA for a few years, said the Chinese were part of the persistent threat we face; they were expanding their efforts; and we were finding it difficult to stop them from being successful. So, it appears they can deny it all they want. They are doing it, and we already know how. This probably annoyed them quite a bit, but it annoys us just as much.

Living in Bad Neighborhoods

When the Internet started to replace television, we probably should have noticed that it was doing more than that. It was changing the way we interact with each other and, among other things, bringing in people who lived in bad neighborhoods. The Chinese are only some of them, but they are the majority of the new people on the Internet. They don't know how to act, but we are learning that the hard way.

The Internet is usually thought of as a neutral place, not good or bad. This is a myth, one that started with a grain of truth, long before the Internet came to be. Before we saw the first mention of the Internet, it was possible to roam around on the computer networks of that day and probably not run into anyone, or anything, that would cause a person grief. It was like a neighborhood where you could leave your doors open at night and people might even come in and walk around the house, but they never took anything or made a mess. There was a kind of strange relationship between the people who owned computers and the people who used them. Mostly, computers were used for good, or from the business side of things, for productivity. Everybody liked that and they felt better about sharing this good for everyone.

By the late '60s, the people who were coming in and walking around started to take things that didn't belong to them. It didn't happen often, but people who operated computers thought it happened often enough that they needed to stop letting everyone in, and started to think about protecting information from anyone who might try to get at it. Some of them were saying "there goes the neighborhood" kinds of things to justify cutting the systems off from each other. Business systems were just connecting to each other for business, but people in those businesses were stealing from each other. A few of them were professional criminals trying to blend in, but not very many. Most of the time, they were just opportunists.

One guy who knew how to program a bank's computers invented a scheme that was pretty clever. He thought that he could slice off a piece of every bank transaction and he could make the piece small enough that nobody would notice. They called that the "salami technique" to make it sound less complicated, but it is not all that easy to do. Still, the money piled up pretty fast, and he got caught, and accounting programs started to round off numbers to eight decimal places—just in case someone tried it again.

Another couple of guys found out that you could go downstairs in an airport and when people upstairs used their American Express to buy checks they could later cash, they could record the electrons that made that happen and play it back to get more of those checks. You don't see those in airports anymore. That is a "playback attack," and we still have some of those around. They just don't always work as well as the first ones did. People are pretty smart at thinking of ways to steal money, and nothing we can do will limit their creativity.

The occasional bank VP would use a computer in his office to make phony transfers to companies they had thought up. As it turned out, a few of them went to jail for it, but not all of them. A guy in Ohio, whose name I have long forgotten, was killed in a plane crash and his wife showed up for the funeral; then his other wives showed up. He had five, in all, and five houses, with wives and kids in them all. He was a computer programmer with a legitimate job somewhere, but he wasn't being paid enough to support that many families. These kinds of things still happen, but back then, they didn't happen very often because networks were pretty safe places to go. What there was of an Internet was just a bunch of networks connected together.

In the early '70s, the Air Force started to worry about security of their computer systems, publishing a report called *The Computer Security Technology Planning Study*. It was Top Secret. James Anderson, who wrote the report, said, "There is little question that contemporary commercially available systems do not provide an adequate defense against malicious threat. Most of these systems are known to have serious design and implementation flaws that can be exploited by individuals with programming access to the system…. The security threat is the demonstrated inability of most contemporary computer systems to provide a sufficiently strong technical defense against a malicious user who is deliberately attempting to penetrate the system for hostile purposes." Today, we could hardly argue with his statement, but things were going to get worse.

IBM had not invented the personal computer, and they had not asked Microsoft for an operating system for it. When they did, anyone with a brain could start playing in my house, and we knew this was not going to work out. The neighborhood started changing and lots of people were moving in around us. They didn't look like people I wanted in the kitchen. We had to start cutting back on the number and types of people we let into our networks. In those days, we called it computer security, but it was mostly just cutting off those connections to other systems we were connected to, and being a little more careful about what our people were allowed to do.

Some people in research were talking about connecting up more networks into a giant ball of networks, using the ideas that had come from ARPANET. The computer whizzes thought that the more of them they could connect, the better the world would be, but not all of us liked the idea. We needed convincing, so they sent evangelists to talk to us. Everyone on earth could have access to business information that they needed to keep our commerce engine humming along to the next millennium or so, they said. We will only have to put data in once and everyone can get to it after that. People can get together and have new ideas flying around like snowflakes. We could all work from home and take care of our kids at the same time. It seemed like a good idea. I liked working from home. They had a convert.

In the early '70s, some ARPANET researchers started talking about computer code that could be used to pass things around from one computer to another without the user being aware of it. There were good reasons for having these kinds of things going on behind the scenes, but there were some bad uses that it might be put to. I remember someone made a Christmas tree virus that brought up a tree whether you wanted it or not, and that probably didn't go over very well in some of the places it showed up in. We took all that to be good fun, but it didn't last long. It turned out there were some really ugly things that could be done with a virus, including wiping out the data that a user stored on his computer. That was bad, but not nearly as bad as it would get.

The last 20 years have not been very much fun, but the last 10 have been the nightmare on Elm Street. There are thousands of viruses and worms that can spread without being connected to another computer, and some new ways that would beat virus scanners and security. There are a hundred ways to attack a wireless system, and the more security they try to get, the less effective they seem to be. Ransomware is a hot topic these days because someone turned loose an attack that hurt hospitals in Great Britain's National Health Care Systems and shut down Renault's assembly lines. It had some impact in 150 countries. It was based on a worm that took advantage of a vulnerability in Microsoft software, encrypted files and sent a message to the user that said "Pay ransom or lose your files." This has recently been attributed to North Korea, as was a well-known heist of money from the national bank of Bangladesh, opening a flood of "North Korea did it" speculation in a number of areas.[9] If this opening shows anything, it will show that we did not know who did it when the event occurred, and that North Korea has been up to a good bit more than anyone knew.

The Internet has spread to a third of the people on earth, and not very many of them can do anything about it either. They are just individuals and most of them don't know what people are doing to them. Those that understand it still can't do much about it. The people who are causing this trouble are organized, protected, and really good. They are making it a bad neighborhood. If you think of this as being only a bunch of hackers trolling around on the Internet making money, you would be looking at this the wrong way. The Chinese are using it to undermine our business structure.

Signs of Decay

Most everyone understands what it means to have people around you who do not understand how they are supposed to act. If you go to a bad neighborhood, you can tell without having any signs that say "Warning." One of my Canadian friends asked me to speak in his classroom at the Washington Navy Yard in the 1980s. Since I wasn't from Washington, he said I could take the Metro and walk down. There is a Metro stop there now, but back then, it was five blocks. The Metro stations are pretty nice and there are lots of people around, even at 7 in the morning. They get coffee, pick up a newspaper and chat occasionally about the Redskins, Wizards or the Capitals hockey team.

As I left the Metro station and crossed to go south to the Navy Yard, I passed under a major freeway and went from pleasant coffee shops and newspaper stands to a war zone. Nothing was open; windows were broken in half of the buildings; and there was glass on the sidewalks, so I was pretty sure the breakage was not that old. It was dark because most of the streetlights were broken. It smelled bad. There was not a soul on the streets anywhere, in any direction I could see. Only a few cars came down the road to the end and they had to dodge glass and miscellaneous rubble, but they did provide some light. I said, "You should not be here," to myself. When you say that to yourself, you have already taken in all the clues and processed them. You don't need a checklist to know it is not a good place.

I stopped at the intersection across from the main gate of the Navy Yard, and a woman was coming across the street towards me. She didn't wait for the walk signal like I did, because she lived downtown where they don't. I said, "Good morning," like we usually did in Richmond. She said, "Are you sure you are in the right place there, son?" I said, "Yes, I'm just going to the Navy Yard, right here." She looked down the street about two blocks and gestured with her head. "The entrance is down there. You can't go in this way. Stay away from those boys on the corner down there and don't be comin' back this way until it gets light." It looked like good advice, and I said, sincerely, "Yes, thank you." I stayed away from the boys on the corner, who did not look like they were going to work or school anytime soon, and took a cab back when I was done.

What is hard for us is that some of us live in neighborhoods that have gotten bad over time. They were beautiful when we moved in but they have started to fall down after years of people moving in and out. People who have lived in them for a long time still think they are nice places because they haven't taken a close look.

A bad neighborhood on the Internet is not so easy to identify. You will not be able to say, "I shouldn't be here," because it won't be obvious that you shouldn't be. The behavior can be identified by discovering and attributing crimes to a source neighborhood, or a collection of them. Attributing means I can say, pretty much for sure, that this address was the place this attack was coming from and not somebody who had taken over that address and was using it in their name. You can see the difficulty there. If I see the Davis Furniture Company attacking me, I can be fairly sure they are not the ones really doing it, unless someone over at Davis has gone off the deep end. It is more likely that the website Davis had his son set up was not quite as safe as he thought and someone captured it. They are just using it to do bad things and Mr. Davis will not know until someone calls him. Even then, he may not know what to do about it, but at least he knows.

People know the place they want to go to by a URL. This show up at the top of your browser, like apples.com, but there are no broken windows or streetlights to let you know that the number it translates to is not a good place to go. The number is an address, unique to that place, but it can be changed faster than moving from one office to another. Some security vendors will sell you a piece of software that will watch these places and tell you to stay away from the ones that are bad. That is a good investment, but hardly foolproof, since they can change them faster than anyone can keep up with. At least, they try.

There is a step in between the typing of a URL and the going to the site that is a translation of that text in the browser to a number. The place this is done is called a Domain Name System (DNS) server. Hackers have taken to attacking these and undermining them so they will send a person to their website, which will look like the one you thought you were going to. A normal human being can't tell the difference, so they get sucked into typing their bankcard number and password, thinking they can get a money transfer done. Those software packages that do security are supposed to be able to protect a person from this type of place, but they can only stop what they know about. Criminals and government hackers know what they are doing and try to adapt their sites to stay ahead of the curve.

Most bad guys will put up false road signs to get me to go their neighborhoods, like a sign directing me to the right when I should be going straight. I have to look closely at the URL to see that it is not what it is supposed to be. If I don't look at it carefully, I will end up in their neighborhood. Once I am there, they can put keystroke loggers on my systems to watch me and steal my credit card info or my personal data. Then they can us my identity. It is a lot like the Stepford Wives, because most people will not know that I am not me anymore. There is more than one me, and there can be hundreds.

Hackers go out of their way to create nice neighborhoods. The Internet is a physical place and a virtual place at the same time, which makes that easier. *Avatar* had the idea right. I can be in a back room in a dive hotel in Kinshasa, but my website shows a stunning crystal front house on a rural mountainside, where only someone rich could live. There are usually children there, and quotes from famous people, implying that this is a great place to buy things. "These unique surroundings provide some of the best investment opportunities a bunch of monks could ever manage for you," they will say. They don't say anything about Kinshasa. They could even be in a government office there and nobody would know.

They can sell pretty photos of naked men and women doing things I had not thought of. This will get my attention, and while I am thinking about how it was possible to do that, they can get quite a bit off my computer and plant software besides. A person should really be careful not to go to places like this, but it is not so easy to ignore them, particularly when you are young. Young people are not as afraid of bad things happening to them, and they value those lessons from pictures of naked people more than I do.

Bad neighborhoods on the Internet move around quite a bit. Other governments, businesses, and hacker groups like to keep an eye on those places, and they set up monitoring to see what they are trying to do and how they are doing it. They produce lists called whitelists and blacklists that indicate how they feel about the different places a person can go out there. Whitelists are places that are allowed and are "good," meaning they are allowed to connect to each other without a lot of questions. Whitelisted places have to keep their systems reasonably secure so they can't be modified to accept bad code from one of the blacklisted sites. The folks who have them don't give them to people like us. Governments keep these to themselves.

Hackers have gotten smarter and can do things now that they didn't know were possible many years ago. The Chinese are using stolen software and code-signing certificates to create their own domains and attract people to them. They have spread those domains far and wide. It is impossible for users to know if the domain is valid or not. Some government-sponsored people were doing the same thing years ago. The first circuit boards that I remember being modified and replaced in a computer system were put in a casino in the '80s, after slot machines were automated. They paid off more than some of the others, but the guys who did it eventually got caught. Hacks like modifying memory sticks and leaving them around an office where potential users can find them are old news in some places. Generations of hackers think they are discovering the newest ways to get in, when they're just rediscovering them. The governments of the world have been doing most of the things hackers do on the Internet since before most of them were born. Only rarely does something new come along.

In 2004, Shadowcrew was one of those. You will not find out very much about this group by wandering around on the Internet except some old stories about two cases. In the first, 19 people were arrested and 9 others were eventually prosecuted. The Justice Department says that the indictment charges that the administrators, moderators, vendors and others involved with Shadowcrew conspired to provide stolen credit card numbers and identity documents through the Shadowcrew marketplace. The difference between them and the people who have come before them is they managed to get into the systems that processed credit for companies. They were not stealing one transaction at a time. They were stealing all of them.

The account numbers and other items were sold by approved vendors that had been granted permission to sell. They had to be screened before they could play in this game. In other words, they got vetted by Shadowcrew to be criminals. Shadowcrew members got at least 1.7 million stolen credit card numbers and caused total losses in excess of $4 million. What a person gets prosecuted for is what the government can prove in court. For everything they can prove, there is a good bit more out there. For its time, it was not a big case at all.

One of those arrested pleaded guilty to acquiring 18 million e-mail addresses with associated usernames, passwords, dates of birth, and other personally identifying information—approximately 60,000 of which included first and last name, gender, address, city, state, country and telephone number. There were 4,000 users of the closed website that was operated by Shadowcrew. They were selling this information to other people, and most of them were not prosecuted in this case.

Albert Gonzalez worked for the government as an informant and helped to break that case. But he was arrested again in 2009 for stealing card numbers from T.J. Maxx, Marshalls, Sports Authority, Target, Barnes and Noble, JC Penny, and 7–11. This time it was 180 million stolen credit, debit, and store card numbers. That is a big number. Now people scramble to buy something that is not a credit or debit card, like my iTunes and Amazon gift cards that can be redeemed online. That way, I can buy things without having a credit card on their computer networks. I know they do the best they can, but that may not be good enough anymore. What Shadowcrew was showing everyone was the inability of the people we buy things from to protect our information, personal data, credit card numbers, e-mails, and other types of things from people who would steal it. The thieves were all over the world, working together, and using the Internet to do it. Shadow Brokers is the group identified as the sellers of software leading to the ransomware attacks. Shadowcrew is not the only one of these groups either, but there are a small number of them because it is a difficult and dangerous business to be in. They were possibly connected to the latest ransomware attack and were said to have been selling attack tools they got from Edward Snowden.[10] We have no idea where this software really comes from, and the speculation carries very few factual accounts to go with it. We are underestimating the amount of harm they do, and overestimating our ability to do anything about it. They can corrupt our ability to use the Internet for commerce, and they are protected.

Statistics and Lies

In 2000, there were fewer than 362 million Internet users. There are three billion now. This is almost ⅓ of everyone on earth, so I'm always skeptical of numbers that big. Afghanistan is supposed to have a million Internet users, and that just doesn't seem possible. Go scan around that country on Google Earth and tell me they have that many.

The problem with Internet growth is math. Dr. John Carroll, who was one of the founding fathers of computer security in Canada, used to say there will be some bad people in every group. They will not follow the rules, and some will be destructive or nasty about it. By the same token, there are some really good people out there who will do the right thing, no matter what, even if they lose by doing it. He said there were about 5 percent of people at either end, and the rest of them were scattered on a bell-shaped curve, who are neither good or bad all the time. So, when you increase any population by a significant number, you increase the bad people who can give everyone trouble, and find a few people interested in weird things like cow manure. It is just math. Using the formula, just simple calculations get you to 105 million really bad people out there. The Chinese use this often as a reason for the increased hacking coming from there. "There are bound to be bad people and we will eventually take care of this problem," they say. What they don't tell you is they take the long view of things, and it will not be in your life-time.

While the number of criminal hackers, and the safety of countries they live in, is increasing, the number of people using the Internet and the things they do there are increasing too. We pay bills, shop, search for knowledge, get service from our governments, communicate, socially network and do business, and the number of those things increases all the time. To hackers, these are called opportunities.

There are a billion and a half Facebook users, and half of them log in every day. Facebook started in 2004, the same year Gonzalez was arrested the first time. Twitter has 175 million users. LinkedIn, which is more business oriented, has 34 million users.

One reason governments don't like to talk about who is on a blacklist is politics. Sometimes, the people on that list are supposed to be our friends. The Nigerians get some of their sites blacklisted because they have always supported, or neglected to act on, a variety of Internet schemes that collect money from the rest of the planet by dubious means like the "I found a bunch of money and all you have to do to claim is send me $250 as a service fee." They did this before there were computers. Although I hardly feel sorry for anyone who falls for this trick after 35 years of its being used, I don't blame anyone but the Nigerian government for allowing it to have gone on that long. They know who is doing it. They know it is successful. They could stop it tomorrow or the next day with a few raids and a couple of dozen people going to jail. They have branched out into credit card fraud now, so they probably are not anxious to stop any of this.

The Russians have the oldest criminal gangs operating on networks. They are best known for child pornography, identity theft, phishing, and computer extortion, the latter having rapidly become something called ransomware. These guys lock up your computer and won't let you have the contents unless you pay a ransom—thus the name. Several Russian gangs are criminal organizations, because they offer services to criminals who need to use a secure attack base. This makes it all the harder to sort out, since it is like offering a garage to multiple gangs that commit robberies. They put their cars there. They do their planning there. They sort through the money. How much do I know about any of this as the garage owner? I could pretend to be deaf, dumb and blind, like the Pinball Wizard, and know very little, an approach that might be good for my health.

Of course, it is not that simple. My little garage is also renting out the planning functions for robberies, doing some of the preparation of banks to be robbed, laundering money, and providing very safe places to work that cannot be seen or heard by police. The police would have a hard time with my Pinball Wizard story.

When the Russian Business Network (RBN), one of the more sophisticated of these groups, started to get some heat from the rest of the world (and maybe the Russian government, since they have been known to cooperate on rare cases in the past), they moved. They tried to hide behind an Italian front company, but that didn't work, and they pulled the plug on some of their operations until the new side of it could be set up. It should not be a surprise that it was in China. Nobody has seen them since, and talked about it. That says they are good at what they do, or they are out of business. Guess which.

These could be isolated examples, but they aren't. One of my professors used to tell us that successful criminals spend as much time at their jobs as you do at yours, and you are not likely to run into one of them that you could recognize. They look normal; they have families and homes; they go to church, sometimes. There are probably more than just the 105 million, because that curve will take in a few hundred million who are just copycats or low-level part-timers; there are a lot of people stealing for a living, and some are really good at it. At some point, we wake up and say, "This is a bad neighborhood." Lots of bad things have already happened to us by that time and we don't know about all of them.

About 1 percent of the hackers of the world are really, really good, so we are looking at roughly a hundred thousand of them. Kevin Mitnick, in *Ghost in the Wires,* tells the story of getting birth records from live and dead people and creating driver's licenses, library cards, and records of all kinds, to be someone else. He kept these identities around in case he needed them. He always had to worry that one of those who were still living would notice that someone else was pretending to be him. He monitored his own phone lines to make sure they were not being traced. He used the information systems to make false records of him. There are a few of these kinds of criminals out there and they don't get caught. They have their own very good security and they watch the people watching them. They make websites to communicate and share their methods and results, but they are a really tight group and not very trusting of one another. They don't trust anyone outside their own community.

Most of them who work for governments will never be caught and will go away if discovered. If they are criminals without government support, law enforcement people can eventually track down some of them, but there are still some who never get caught. We tend to measure crime by people who do. It may be years before some of this type of crime starts showing up in the records of credit card companies and government files. In the same way, government hackers do not get caught. They are good at what they do and can melt away if even a whisper of discovery is heard.

As the Internet Churns

Internet crime was growing before the Internet got to be as big as it is today, but there are some things that make it a more dangerous than it was even 10 years ago. Using the Internet to attack us is part of the national strategies of a growing number of countries. Our new neighbors on the Internet are people we don't get along with very well; they don't know how to act; they are not like us.

Most people added to the Internet since 2000 are not North American. That is not an obvious thing to most users of it, since nationality or cultural background does not show up anywhere. If I see a .ru or a .cn after a URL, I know that is Russia or China, but it doesn't say anything about what nationality the person operating the site really is. That just says where the site is registered, not where the operators are physically located. The person who manages the site could be a Latvian on a tourist visa in New York. When RBN went to China, it was looking for a safe place to not be seen.

Most new Internet users are from China or India. The United States is third. China has twice as many users on the Internet as the U.S. has in total population. That is not happening without China's interest in having their population on the Internet. They encourage it. We should probably remember that China is not our friend, and we are building up our military in the Pacific and cutting it a lot of other places, just to make sure they know how we feel.

A friend of mine lived in China, working for a major company in Europe, and she gave us an example of a culture difference that she noticed in the first week there. She tried driving herself, but only once. On her first day, she came to a railroad track, noticing that a train was coming and most of the people stopped. Some didn't, but that happens everywhere, with people trying to get through before the train got there. But while she was sitting there, she noticed that people were starting to drive up next to her in the lanes that would normally be for cars coming in the other direction. Bicycles and scooters were taking up spaces between the cars. She also noticed they were doing the same thing on the other side of the train. When the train was gone two opposing forces were taking each other on in a game of chicken, all trying to funnel themselves back into normal traffic. She couldn't deal with it and hired a driver.

They are certainly culturally different in many ways, and some of these are important to how they think about the use of the Internet. They are poor. Half of the population lives in cities. They are overcrowded and live in spaces that we would normally call built-in closets. This isn't a crime or a reason for looking down on someone, but it is a reason for being careful. I wonder how people with incomes just over $8,600, on average, can afford Internet service. It would have to be free for people to afford it. The average salary of people in China is about $8,600 a year and only 2 percent of them earn enough to pay taxes.[11] The upside for them is they spend a good deal less on basic essentials.

China is officially atheist. You might ask yourself if religious affiliation has anything to do with our interaction with Chinese on the Internet. Some coworkers of mine went to a country in China's sphere of influence and they went as missionaries. They told us before they left that we should never refer to their status in any written correspondence or e-mail. We should never mention the church that they belong to. We should always refer to them as teachers. They knew their mail would be read. They knew the country they were going to was not very tolerant of Christians. They were going to live in the mountains a long way from any civilization as we know it. When we started to think about what it was like to live like this, it made us appreciate living where we do. There are quite a few countries that could find ways to make a person's life very unpleasant for something they said on the Internet, and it is not how most of us want to have to live.

China is always a mystery, but there sure are a lot of people there. They have tried to use population controls of one child per family that certainly would not go over very well in most other countries. They go to great ends to control their population, particularly those who may not agree with how the government is being run.

The Internet is bringing people together who are quite diverse, and while that normally wouldn't matter, there are times when it does. I grew up in a time when do-gooders like me were going to school to learn how to eliminate hunger and get world peace by helping everybody get along. It takes some time to figure out that people who are really, really poor are different from middle-class America. We called that the "culture of poverty" to try to sum it all up.

Besides China, some of the biggest Internet user gains for their population size were in Russia, Indonesia, Brazil, Nigeria, Iran, Turkey, Mexico, the Philippines, Vietnam, and Argentina. In a way, the Internet has brought us closer to these people, sort of like cheap air travel has, but there are some of these that I don't want to be closer to. I don't use cheap airfares to fly to Russia or Iran either. There is all of the usual benefit of "international understanding" that some will talk about, but the understanding seems to come from us and not them.

Some of them have a history of not behaving on the Internet, or anywhere else for that matter. So besides being poor, they cause trouble. China now leads the world in computer hacking sites and general troublemaking. China, particularly the army, shows up more and more and has become the center of attention for the U.S. intelligence community, because the Chinese have been hacking almost anything that has an Internet connection and stealing everything they can. The estimates for theft of businesses' intellectual property are $1 trillion every year. The Commission on the Theft of American Intellectual Property, run by Jon Huntsman, who was ambassador to China from 2009 to 2011, says the majority of the theft is by China. The Commission Report says there are not sufficient enforcement mechanisms in the Federal Trade Commission, the CFIUS, and the U.S. Patent and Trademark Office to allow a company to stop the manufacture of goods based on stolen intellectual property, so there are few remedies once the data is stolen.[12]

There are stories on the Chinese hacking U.S. satellites, planting software in our electricity grid, and hacking our defense and businesses at such a rate that they got the attention of the president—who they also hacked during the election. They certainly have been busy, but they are just the newest bully on the block. Russia has done most of the hacking over the past 10 years or so, very similar to what China is doing now.

The one difference with the Chinese is simple. They are hacking everyone in the free world and they don't mind us knowing. They are pretty open about it, but they deny everything. I have enemies like this. They will deny, right to my face, something I saw them do. I have recorded evidence. I have my own observations, but they deny it anyway. I switch them over to my own blacklist, which I keep in my head, and I watch them closely after that. I take more notes about what they do, and research them. I never trust people who can lie to my face.

Iran is new to the business but has a couple of well-known hackers who have done some good technical attacks against security firms. Iran is not our friend, and that leads me to say that it should be watched. Iran and China share a good bit of their networks, and neither of them is our friend. They are probably sharing their hacker knowledge, but it would be hard to prove that one way or another. They are not going to say much about it if they are.

We used to track hackers regularly going through Brazil, so they are either really poor at security down there, or they are allowing it to happen. Maybe both. I have never seen a hacker from Vietnam, though there are some. Vietnam and China went at it in June of 2011 during an argument over some islands that both of them claimed. They hacked each other's websites and tried to do a few other things, but it seemed to die out soon. Neither one of them likes to talk about that, so there is not much written about the types of attacks that were really being used. Behind those website attacks, there was probably some interesting information warfare happening too. Both of them are getting good experience, so this is not going to go away unless Vietnam decides it does not need those islands it is dredging out.

Internet use is increasing in places with growing populations; and some of those are countries that don't like us very much. Only a third of Chinese have a favorable view of the United States. This doesn't always mean war, of course, but it is certainly a measure of how predisposed people are to thinking about it. They like Japan even less (71 percent have negative views), and that must give the Japanese quite a bit to think about as the Chinese increase their military missile and combat forces. They are closer to Japan, and some of them remember the Japanese invasion of their country before World War II.

Pakistan could do more damage than any of the rest of them because they have many inroads into U.S. companies operating in their country, and they are not exactly friends of ours. I would like to know why so many of our credit agencies, medical records, and computer center help desk companies think they can operate from there. These are really, really sensitive operations to have in a country that was keeping Osama safe, helps other groups get nuclear weapons, and loves the Chinese. The only two countries with majority negative opinions of the USA were Russia and Pakistan.

We are dragging people online who know next to nothing about computers but want to do things on the Internet anyway. As my engineers used to tell me, you cannot engineer-out "stupid." At least 500 times, companies have put out announcements telling people there was a phishing attack directed against them. This sort of e-mail has an attachment with the headline "New Company Logo." Clicking on the attachment will cause a computer to explode and splatter pixels all over you, along with some other such thing that will be more technical. A normal person would think twice about clicking on that attachment, but some people reading their e-mail are not normal. A few of them will not have read the corporate e-mail since they were on vacation and start with the most current trying to catch up. A few are curious about what pixel splatter actually looks like. A few will think it applies to everyone except those few people in corporate management places where people take care of them. They actually believe that nothing happens to exempt leaders. Those that do click on it will be rewarded with new faith because nothing does happen. Nothing they can see.

Our youth go to those sites in Russia and Eastern Europe that advertise with naked women in various poses of artistic merit. I loaned a computer to one of my neighbor's kids and it came back with a few of these, plus loads of other problems that suggested he had been going to quite a few parts of the world that are bad neighborhoods. He wasn't old enough to have a credit card to pay for the videos, but I talked to him about how that might work for him when he gets one. He trusted people on the Internet because he didn't know any better. Those engineers didn't know any better either. Besides the other things they were doing, we did catch a few engineers going to sites that had very little artistic merit, and we had one corporate VP selling 2,000 images a day from his government-provided computer network. For almost anything a person will do, there is someone who will help them, but there is going to be cost for that.

Epsilon did a bulk collection of names and e-mail addresses so they could send out ads and notices to customers of their clients. I know this because I read the news, but also because some of the people like Best Buy, Staples, Verizon, Ritz Carlton and the *Wall Street Journal* have sent me e-mails to say that those addresses were stolen. Somebody has my e-mail from all of these places. Press speculation is that information might be used to craft letters trying to get more information from me, but I stopped looking at these e-mails right away. They can just keep sending the ads out with embedded scripting in them to get access to my system. They think I will still open those attachments. Quite a few people will, because it is too hard to see the difference between one ad and another that looks just like it should.

The folks at Epsilon reported it, but it could have started months earlier. Maybe some of the ads I have been getting were already sent out with those little bits of code embedded in them. Now the hackers know I shop at Best Buy and Staples, stay now and again at the Ritz, and subscribe to the *Wall Street Journal*. I have already received notice that my subscription needed to be renewed at the *Journal*, and it wasn't the *Journal* telling me that. Whoever bought this information can send me letters to make me think my last credit card transaction was kicked back or asking to verify the number because my subscription has lapsed (that actually did happen, and ours was not expiring until next year). They can look for credit card numbers and do quite a bit of damage before someone can stop them, depending on how much they are able to collect and correlate.

This is all the more reason for businesses to do better security with data they keep. Companies such as RSA, Lockheed, Northrop Grumman, Target, Sports Authority, BJ's, T.J. Maxx, and Stratfor, plus hundreds of small businesses all have been hacked. It wasn't a surprise to anyone in computer security.

One of my teams was doing a survey of a small company with venture capital funding. They were going into a new line of business for them, storage of online data for external customers. The venture capital firm hired our company to check out their security before giving them any more money. It's a good idea.

The company basically had no security that would allow them to do business on the Internet. We sent our team out twice because we didn't believe the first set of results. The capital was withheld, and the company never went public. It would have taken only a few minutes to get control of that site and all of the information stored there. When companies can't operate on the Internet without losing some of the most sensitive data they have, such as customer lists, social security numbers, credit card numbers, and employee information, then they shouldn't be operating on the Internet. But many still are. This is because businesses see risk differently than we do. If they lose my credit card information, they are only liable for a small amount of the losses that can come from that. I suffer much more than they do. As long as that is true, the burden of risk will always be on the small user and not the companies putting all that information together and losing it.

The Federal Trade Commission estimates that there are 8 million cases of identity theft in the U.S. every year, and this is partly where they come from. The crime that results from this is not always visible, since sometimes they don't even know their identity is stolen. The Internet makes it easy, and even advantageous, to be more than one person at once. A person can be 25 people and use the credit cards of all of them in the same day. The companies that lose the information are not the ones paying the bills.

New Internet users are not people of our culture. I don't mean of our nationality, because that isn't enough. These are not Western cultures. That may not matter to some, but it should. There are some really, really bad people on the Internet who peddle child pornography that will make you sick to your stomach, and make those Russians nudes look like children's book material. But nobody can reach them where they live because they have protection. They sometimes capture videos of what they do to these kids and sell them to make money. There seems to be a never-ending supply of people to buy this stuff, and neither side of those transactions is the kind of person I want to "like" on Facebook.

China hacks us from multiple places and across business and government targets. They are very successful at both getting in and stealing our secrets, but given the demonstrated state of the art of defense, that may be easier than it should be. They make the winning bids; they get to compete everywhere in the world. These are the morals of international business newly defined by China. They have managed to make up their own rules and get others to follow them, leaving us with the unpleasant choice of adopting theirs or standing with our own. I'm for playing by theirs, but it is not as easy thing to do.

Crime is just one risk that we face in a bad neighborhood, but this risk is going up pretty fast and the users of the Internet are not taking stock of it. We don't have many measures to say when this risk is too great. I can't decide something like that for myself.

What makes that a dangerous thing to say out loud is that our commerce depends on our ability to keep computers secure, and to say there is something wrong with that security makes businesses on the Internet jump. They don't like anyone saying it. Their customers have to believe it is *reasonably* secure, a term invented by people who were looking at risk at a time when there wasn't very much. What the Chinese are trying to do is undermine the ability to do trade safely on the Internet. They undermine our commerce to make war, while the Russians interfere in our elections.

When U.S. government investigators looked at Facebook after the 2016 national election, it found some of the accounts created to foster events or influence Facebook users were Russians working for a company called Internet Research Agency LLC. Those Russian sponsored accounts were used to support political stances for both the Democrat and Republican parties. The used networks in the United States to make tracking their country of origin more difficult. Twelve individuals are named in the indictment by the U.S. Justice Department.[13]

The indictments of Russians, in Russia, will not deter anyone from carrying out the kind of operations that were described by them in their own internal communications. Normally, these are intelligence intercepts and we would never see them because they would be state secrets, but they were used in court to establish a case that could be brought to trial. What is in that description is a window into the way Russian Information War is intended to work. *The New Cyberwar* described what they had done in the Ukraine to disrupt the Ukraine government and aide the southern rebels in fighting the established regime. Those operations are called "reflexive control" means controlling the narrative around a set of events to favor positions that are best suited to Russian objectives. The Russians believe that leaders can be persuaded through selected information (not necessarily truthful information) to accept a path that is suited to Russian objectives.[14] Secrecy and denial are major parts of Russian involvement, characteristics of covert operations. In covert operations, plausible denial is essential to the outcome. A country must be able to say, "it wasn't me" when questioned about their involvement in these kinds of activities. This particular operation was current and directed at the general election for the President of the United States.

Those indicted in the Russian operations to influence the U.S. election are both companies and individuals. The companies are: *The Internet Research Agency (IRA) LLC,* [also known as MediaSintez LLC, Glavset LLC, Mixinfo LLC, Aziut LLC, or Novinfo LLC,] *Concord Management and Consulting Company* and *Concord Catering.* The individuals were all Russians, most working for the Internet Research Agency, and all living in Russia. Vladimir Putin says he will never release these individuals for trial. The important thing is that the individuals are identified by name, and as an indictment would infer, there is sufficient evidence to make a case to try them. If they ever venture into a jurisdiction that has extradition with the United States, any of those people can be arrested and transferred to the U.S. for trial. Until then, we will not know if the evidence is sufficient to convict them.

What the Russian IRA was doing is not exactly new, but it was an updated approach using modern media. The Russians have been meddling in U.S. elections since 1982, and half-heartedly tried to prevent Reagans nomination as the Republican candidate in 1976. But in 1982, then Soviet Premier Yuri Andropov, (just as Vladimir Putin today), a former KGB officer, decided that all KGB officers should participate in active measures aimed at preventing Ronald Reagan from having a second term in office. It was the KGB's highest priority during the last months of Andropov's rule. The Soviets decided that any candidate was preferable to Reagan, and their actions should promote a slogan, *"Reagan Means War."*[15] At the same time, they directed five active measures (themes) against the Reagan campaign, two years away:

1. Discredit his foreign policy. Characterize his policies as militaristic adventurism. Identify Reagan as personally responsible for the arms race between Russia and the United States. Emphasize his support for repressive regimes around the world. Describe his administration's attempts to disrupt or destroy national liberation movements. Show how he has created tension with NATO allies.
2. In domestic politics, describe Reagan as discriminating against ethnic minorities, having a corrupt administration, and being subservient to a military-industrial complex.[16]

Although they devoted the remainder of the two years to these activities, Reagan won a landslide victory anyway. But the Russians have learned quite a bit from their mistakes in trying to influence an election process that is a little unpredictable, even to those in the U.S. familiar with it.

The 2018 indictment leaves out many facts that were doubtlessly presented to the Grand Jury, but briefings to several Congressional Committees have filled in some of those. Many of the facts that became evidence are still state secrets and will not likely be presented in open court.

The indictment makes it clear that it is illegal in the United States to make "...certain expenditures or financial disbursements for the purpose of influencing federal elections." The laws bar "agents of any foreign entity from engaging in political activities within the United States without first registering with the Attorney General. And U.S. law requires certain foreign national seeking entry to the United States to obtain a visa by providing truthful and accurate information to the government." One could conclude from these referenced behaviors that the Justice Department believes the Russians attempted to influence the federal elections and in doing so, sought visas under false pretenses and did not register with the Attorney General. The indictment says the Russian employees "knowingly and intentionally conspired with each other to defraud the United States by impairing, obstructing, and defeating the lawful functions of the government through fraud and deceit for the purpose of interfering with the U.S. political and electoral processes, including the [P]residential election of 2016."

Money for these operations came from the "Concord companies" and not directly from the Russian government. In any covert operation of this type, funds must be seen to come from a source unconnected with the government sponsoring the activity. What they did with that money has become a source of controversy much greater than the impact of those operations. They paid for Russians to develop on-line personas, pretending to be U.S. persons "operating social media pages and groups designed to attract U.S. audience" participation.

8

It Is Just Business

Most of the time we do not care who makes the equipment we use, but we do care about the utility of the thing we are buying. We might want to rethink that in the case of computers made in China. What makes this situation of concern is the willingness of the Communist government to use its products to acquire information about the people who use them. This is not the same kind of debate that occurs in the free world about how far a government may go to get information about its citizens. This is about spying on people inside and outside another country. The irony is the complaint, levied by Edward Snowden, that the U.S. spies on almost everyone in the world, when China is holding its own in doing exactly the same thing without getting the criticism it deserves. Maybe that is part of the reason Snowden went first to Hong Kong instead of his current home, Russia. China uses these devices to spy, and it does so as part of its national policy. In spite of recent global concerns about encryption of cell phones and computers, nobody escapes being monitored, and this monitoring is almost impossible to detect.

A number of years ago, I was in a group that was looking at how computers were being equipped with monitoring equipment when travelling through other countries. Some of these computers travelled in bulk, and some came into the U.S. from individuals travelling to another country. In one case, we had an informant in another country tell us that some of the computers coming from his country had been tampered with and monitoring equipment was built into a circuit board of a particular brand of computer. Our technical specialists took the boards apart and examined them, finding nothing. They went back to the informant, thinking he had made a mistake on the batch that was affected. The informant was insistent. The technicians went back in and finally found what they were looking for. Something that enables another country to monitor a computer can be put into the hardware or software of that computer and it will look and function like a normal device. Even when these devices are disclosed by internal informants, they are almost impossible to find. China would say we have no reason to question their products, but that is only true for consumers who do not pay attention to what has already been discovered.

Probably the best known use of technology to spy on individuals was the Chinese Green Dam Youth Escort. In July 2009, the Ministry of Industry and Information Technology (MIIT) issued a letter requiring computer manufacturers to pre-install this software on computers made in China or imported into the country. The government maintains that Green Dam was a legitimate tool to reduce pornography on the Internet, thus protecting its youth. This was the same rationale the Chinese used with Google to try to force it to restrict access to search results that China did not favor. When that effort failed to convince Google, Chinese hackers went after accounts held by dissidents with Google accounts. We would hardly believe that the hackers were going after dissidents to keep pornography out of the hands of Chinese children. We should have the same reasoning with Green Dam, since it filtered politically sensitive and religious websites as much as pornographic ones. The OpenNet Initiative published a report on Green Dam that said:

> The version of the Green Dam software that we tested, when operating under its default settings, is far more intrusive than any other content control software we have reviewed. Not only does it block access to a wide range of web sites based on keywords and image processing, including porn, gaming, gay content, religious sites and political themes, it actively monitors individual computer behavior, such that a wide range of programs including word processing and email can be suddenly terminated if content algorithm detects inappropriate speech. The program installs components deep into the kernel of the computer operating system in order to enable this application layer monitoring. The operation of the software is highly unpredictable and disrupts computer activity far beyond the blocking of websites.[1]

The U.S., Japan, and a number of others complained about Green Dam. Since most computers today are made in China or assembled from parts made in China, the reach of such a policy is broad and dangerous. It shows the willingness of the Chinese government to impose requirements on products and services that have impacts far beyond their own country's border, even to the point of products made outside China. It also did not make exceptions for products made in China for export to other countries. We have no way of knowing what other types of requirements may have been levied on the electronics industry, and how those may further the ability to monitor users of Chinese products. Those are all state secrets in China.

The Green Dam complaint never actually got to the WTO, because China backed off of the requirement rather than face the WTO action. China, however, allows the voluntary installation of Green Dam on computers there, allowing millions to be so equipped. In the past few years, China has found a new way of getting even more information from its users—and ours.

In 2015, some seemingly unrelated findings by various security companies point to Chinese-made computers having software installed that undermines the capability of securely transmitting to secured websites. These incidents relate to something called Transmission Layer Security (TLS), which is the gold standard for security of transmissions from a computer to a secure website. What caused security companies to start paying attention to Chinese involvement in TLS was a piece of malware, Superfish, that was pre-installed on Lenovo computers, the largest seller in the world. A *Hacker News* article at that time described the effects this way: "Superfish uses a technique known as 'SSL hijacking,' [and] appears to be a framework bought in from a third company, Komodia, according to a blog post written by Matt Richard, a threats researcher on the Facebook security team. The technique has ability to bypass Secure Sockets Layer (SSL) protections by modifying the network stack of computers that run its underlying code.... Komodia installs a self-signed root CA certificate that allows the library to intercept and decrypt encrypted connections from any HTTPS-protected website on the Internet. The company's SSL Decoder like Superfish and other programs are present in numerous other products as well."[2]

Superfish allowed the installation of features that undid the security of Transport Layer Security (TLS) by compromising the root encryption system that made it secure. Lenovo maintained that this software was used to inject adware, the advertisements that we see when we go to a website, but it was more than just an adware vehicle. Those who knew it was there could use the flaws it created to get into those computers. Lenovo at first denied that this kind of flaw should be taken seriously. But it quickly found out that when security companies lock onto something, they tend to look more closely. It was far more serious than Lenovo first claimed.

The first thing the security industry found was the use of Komodia code libraries put the same code into other applications. Since the Chinese develop so much of the world's software, they produce development kits that include the same code. This can be seen as "accidental" or it can be seen as part of an overall strategy to reduce the effectiveness of TLS. When it is just Lenovo computers, it is easier to claim an accident of developers; but it wasn't just Lenovo.

Dell computers were found to have a similar vulnerability for those with the Windows operating system. The U.S. Cert, the national authority that keeps track of vulnerabilities and tries to get vendors to correct them, said this:

> Dell Foundation Services (DFS) is a remote support component that is pre-installed on some Dell systems. DFS installs a trusted root certificate (eDellRoot) that includes the private key. This certificate was first installed in August 2015. An attacker can gen-

erate certificates signed by the eDellRoot Certificate Authority (CA). Systems that trusts the eDellRoot CA will trust any certificate issued by the CA. An attacker can impersonate web sites and other services, sign software and email messages, and decrypt network traffic and other data. Common attack scenarios include impersonating a web site, performing a (Man-in-the-Middle) MITM attack to decrypt HTTPS traffic, and installing malicious software.[3]

We could discount both Dell and Lenovo as accidents until Google, in April 2015, quietly announced it would stop accepting certificates issued by the Chinese Internet Network Information Center (CINIC). This hardly raised a ripple in the security community because the underlying problem with the certificates was technically solved by an additional security feature called public-key pinning, which Google's browsers use. The announcement was almost imperceptible because it came out as an entry in Google's security blog. But there was a good deal more to it than its casual release would indicate.

Google discovered that some of its certificates that protected its domain and Gmail servers were not very secure. It conducted a joint investigation with Chinese authorities over issuance of Gmail and domain certs issued by Mideast Communications Systems (MCS), Cairo, Egypt. These certificates were installed in such a way that "rather than keep the private key in a suitable hardware security module, MCS installed it in a man-in-the-middle proxy. These devices intercept secure connections by masquerading as the intended destination and are sometimes used by companies to intercept their employees' secure traffic for monitoring or legal reasons."[4]

Google, known as the technical leader on the Internet, has quite a bit of experience with China. It said that both Google and the CINIC believed that other certificates were not issued by other companies, and the CINIC said it would do better in the future. That sounds like the kind of press release that would take place after a truce had been called, but the truce did not result in the acceptance of other certs that were issued by CINIC. That is a clear indication that the whole truth of this is missing in public sources. Google would not take this kind of action without knowing what the consequences would be if it were not accurate in describing what was done and why. Perhaps there needs to be room for the smoke to settle.

Starting in 2014, CitizenLab, at the University of Toronto, began to look at the privacy characteristics of a few widely known browsers. It based some of its research on the findings at Stanford and Carnegie Mellon that looked at Safari, used on Apple computers, and Firefox, which is widely deployed. There were some differences in security of privacy data in these browsers, but there were many more differences in the browsers produced in China, and the effects go beyond the borders of that country.[5]

The most widely used Android browser in China and India is the UC Browser made by Umeng, a subsidiary of Alibaba. When a user has this browser "any network operator or in-path actor on the network can acquire a user's personally identifiable information" of various types, some of which have little to do with the user's browser experience. These browsers collect some data that seems totally irrelevant to why a user would do an Internet search, including these items:

- International Mobile Subscriber Identity (IMSI). This is a unique identifier that defines a subscriber, including the country and mobile network to which the subscriber belongs.
- Wi-Fi Media Access Control (MAC) address. A unique identifier that is used to identify a device to a Wi-Fi access point that restricts access.
- User geolocation data, including longitude/latitude and street name.
- Data about nearby cell towers and Wi-Fi access points.
- The Windows version of Baidu's browser also transmits the hard drive serial number, model, network MAC address, and CPU model number.

Neither the Windows nor Android versions of Baidu's browser protect software updates with code signatures, meaning an in-path malicious actor could cause the application to download and execute arbitrary code, representing a significant security risk.

We could argue, as China does, that this kind of thing happens in code development and it is impossible to detect or prevent, and there is no evidence this data that is collected might actually used for anything. What if Edward Snowden had said that NSA does collect quite a bit of data but does not use it for anything? Would that argument have stood up in public debate? However logical it sounds when flowing out of government offices, it isn't very believable. Collecting this breadth of data suggests it must be used for something more than determining where users search. Why does anyone looking for search data need the serial number of the users' hard drives?

The data being collected can be used by hackers to specifically target an individual, narrow an attack to a group of specific individuals, or identify where and how they communicate with others on the Internet. The attacks can locate and track movements of those groups, down to a specific individual in that group. It can locate and target phishing attacks that will allow access to a specific computer, and they know where to find that computer. It makes attacks much easier to pinpoint and check for successful penetration. We are tempted to conclude that this data collection is only good for spying on other people and making sure that their networks are known, tracked and vulnerable. By the number and variety of methods used, China can have a capability that matches anything Edward Snowden accuses the U.S. of doing. They can locate and target anyone who uses their products. And as in the case described by Snowden with NSA, it does not matter if they use the data they collect or not. The issue is that they collect it.

Given the number of companies involved, and the similarity of the attacks against the infrastructure of the Internet, there is a clear indication that there are one or more secret directives from the central government that require vendors to collect this kind of information for use. The companies themselves would not all arrive at the same point of collection without a general agreement about what they would be collecting and why. They know what is required by the government, but secrecy rules prevent them from telling. CitizenLab was certainly thinking of that when they asked Alibaba if it was direction by the central government to collect this kind of data. Alibaba declined to answer the question. If Chinese companies are being asked to collect the data, it would be a state secret that could not be discussed with an outside organization.

What these techniques provide to China is the potential to collect large volumes of material from almost anyone. These are just vehicles that are used to acquire information, and although they are related types of data, that does not mean it would be possible to collect and analyze the quantities that would be collected. Part of the argument China would use would be that it would be impossible to monitor everyone everywhere on the Internet. And that is true.

Gartner says that the total number of devices sold including PCs, tablets, ultra-mobiles and mobile phones will reach 2.5 billion units this year, a 7.6 percent increase from last year.[6] If even half of them could be accessed by the Chinese, they would have to monitor a billion new systems every year. As a practical matter it is impossible to collect, store and analyze all data produced on that many computers, but it gives them the capability to monitor any one of them they choose.

Every month, Google processes more than 100 billion queries, and they are only one of many search engines.[7] There are over 130 billion e-mails sent every day.[8] There are also online chats, video exchanges, messaging and online postings that produce volumes of material every day. Word processing documents, spreadsheets and presentations are created by the millions. It might be possible to collect all of that, but it becomes impossible to search, catalog, and make sense of that much. One of my analysts computed the size of computers required to analyze all of the intrusion detection data our organization was capable of collecting. When he first briefed on his findings, he started by quipping, "It would need a computer the size of the state of Ohio."

No intelligence service can analyze all the information it manages to get. But it is the impossibility of monitoring everyone that makes it necessary to be able to monitor anyone.

China appears to believe that whatever it has to do to monitor its own population is acceptable, even if it allows monitoring of, and interference with, systems outside its own country. The argument is similar to, but the opposite of, the one the National Security Agency has made about collecting and analyzing telephone calls outside the U.S. When using collection methods that suck up data about a wide swath of people, it might be difficult to say who among them is a citizen of any given country. The NSA is prohibited from monitoring U.S. citizens inside the U.S., but people such as congressional leaders and businessmen travel overseas and have business conversations with many people who are not citizens of the U.S. The NSA sets up elaborate procedures to deal with the data that is collected on people who are U.S. citizens who may be inadvertently monitored in those conversations. During the U.S. national election more information was published about the process of "masking" and "unmasking" U.S. citizens than at any time in the past. The motives of Susan Rice, President Obama's national security advisor, were questioned by congressional members in both major political parties because she unmasked U.S. citizens; i.e., she caused the NSA to disclose to her the names of U.S. citizens in those intercepts. While this is not a common procedure for most investigations, the actions in this case involved members of the Donald Trump election team. They were said to be having conversations with Russians that were potentially of national security interest. China uses the same logic in justifying why it monitors its citizens.

China says it is undermining portions of the security controls of the Internet so it can monitor its own citizens and potential terrorists. It does this for their own good, i.e., to satisfy national security interests. In Xinjiang, there have been a number of incidents over the past few years that the international press picked up in spite of China's best attempts to control them. Most of these are mild by terrorism standards, although to a victim of an attack there is very little difference between being struck by a meat cleaver or having a bomb blow up in the midst of people shopping. We cannot dispute the need for stopping the attacks, but we can characterize what is being done in the name of that action as going well beyond what is necessary to stop terrorism. They simply use terrorism as an excuse.

There are few better examples than that of the exploitation campaigns against members of the Tibetan community, journalists and human rights workers in Hong Kong and Taiwan.[9] The ArborSert report identified a series of attacks with similar profiles using "bait documents" extracted from the

Internet. Titles like "Human Rights Situation in Tibet" and "Prediction of the 2016 Presidential Election" (referring to Myanmar's general election), and specifically targeted references to subjects known to be of interest to certain leaders of groups, were poisoned with malware that would give the Chinese access to accounts of anyone opening certain files. The AborSert report says,

> This recent activity matches pre-existing targeting patterns towards the "Five Poisons"—organizations and individuals associated with perceived threats to Chinese government rule: Uyghurs, Tibetans, Falun Gong, members of the democracy movement and advocates for an independent Taiwan. This targeting scheme, along with various malware artifacts and associated metadata, suggest that the threat actors herein have a Chinese nexus.

Combined with the new laws authorizing police monitoring and control of human rights groups and other NGOs in China, the stepped-up Internet spying is consistent with other attempts to control dissent outside China, while it clamps down on similar groups inside.

The main effort is an attempt to censor and limit communications between people who might be dissidents, while potentially contributing to the interception and analysis of intelligence about terrorist plans. In order to do both, it needs a disciplined national information technology infrastructure, something few other countries have.

The Russians are slowly moving in the same direction, but starting from behind. The Russians tried to automate monitoring of the Internet but had a disorganized approach that left the funded research unused or discounted. It was not until 2011 that the equipment and focus made the placement of monitoring devices at ISPs a practical reality.[10] Before that time, the devices were mandated but not coordinated or integrated; now the problems are centered on the ability to analyze the volume of data without requiring large numbers of people from each company operating there. The Russians may be premature in some of their control procedures, because the Chinese make these types of controls look easy.

Andrei Soldatov, co-author of *The Red Web*, tells a short history of the attempt to collect and store huge amounts of data from the Internet. It has been much harder than the Russians thought it would be.[11] For one thing, the estimates for storage by various service providers are nearly 60 million terabytes of storage. By comparison, Google adds 24 terabytes a day to its YouTube service and is geared to handling that much data over time. Because the volume is so massive, the Russians have some difficulty enforcing their own policies, and they are not exactly getting the support of foreign businesses, who do not believe the Russians can be serious about the policy they are asking to be enforced.

To manage these large volumes, Russia has turned to China, in fact to the master himself, Fang Binxing.[12] Binxing is the architect of the Great Firewall, in modern times. The Chinese suggested "white lists," the term for a standard industry practice that labels interconnected sites as either friend or potential trouble. There is also a "black list" for the latter. They undoubtedly suggested much more, since those two things will not make a firewall viable. That is the technical side of information management, but the Russians have used a group called the Safe Internet League to enlist 5,000 volunteers to search for things they would like to get off of the Internet in Russia. That is the beginning of censorship, the real purpose in the Great Firewall of China that they are trying to emulate. They still have a long way to go when comparing their capability to China, and they know it.

In the U.S. carriers such as AT&T, Verizon, Sprint and T-Mobile are regulated, but they are independently owned and operated. The U.S. government does try to regulate their business and, to some extent, controls their content—though viewers with small children might argue they do not do enough. But from the standpoint of censorship, there is none. Nearly all the censorship of content is related to criminal offenses of users, especially pornography and financial crimes. The rise of Facebook Live and a few online crimes shown there, including murder and kidnapping, have increased the pressure on service providers to do more. Curiously, both the Russians and Chinese claim their censorship is related to exactly the same thing.

China seems obsessed with media control, from what kind of music the Rolling Stones played on their trips to Hong Kong to what movies are shown on commercial television. They control the TV news, radio, print media, and film. More importantly, their telecommunications are managed by state-run telecommunications companies and a layer of bureaucracy that is enough to discourage anyone from wanting to be part of it. Yukyung Yeo writes,

> As for the internal governing structures, even though the MII is the formally designated regulator, on top of the MII there are the party and two powerful comprehensive state institutions: The National Development and Reform Commission (NDRC) and the State-owned Asset Supervision and Administration Commission (SASAC).... Unlike the Anglo-American independent regulator model, the MII's regulation is constrained by these top party-state institutions as long as it remains a government agency, for it is institutionally subordinated to the NDRC. Moreover, most industrial policies and regulations drafted by the MII should be reviewed by the NDRC before the State Council's final stamp.[13]

In the case of China, the control of communications is essential to two things: the use of the infrastructure for government sponsored espionage, and the discovery of dissidents who might persuade others to revolt against the established government. Mike McConnell, a former director of NSA, says there are 100,000 hackers working for China, hacking mostly U.S. businesses to collect secrets from them.[14] We might debate the numbers cited, but there is no longer a doubt that China hacks large numbers of U.S. businesses with the intent of stealing proprietary information and plows that back into its economy. McConnell believes they have managed to get into almost every company in the U.S. Certainly, because of China's controlled infrastructure, hacking from or to China has more of a chance of being detected than does hacking from the U.S. The Chinese know two things: who is hacking from their country, and who is hacking into China from other countries. What state-sponsored espionage in China has—that the U.S. and few other countries have—is a secure, monitored network to attack and defend from.

The basic approach by China's internal regulators is to not allow technologies that can hide secrets. Inside China, virtual private networks are prohibited to make it easier to look into corporate networks.[15] The central government requires software used for encryption and internal security to be provided to the government, but publicly says it does that only rarely.

Hackers do not hack inside China without being detected because the architecture includes three components that are not found in any other national network: the Great Cannon, the Great Firewall, and the Golden Shield. These exist more as ideas than physical devices. Those ideas are implemented in various parts of the network architecture and those parts function as a single device would on a smaller network. A home computer has its own firewall, but it is smaller and less capable than a firewall used in a corporate network. Even if we duplicated a small home firewall throughout a large network, the effect is not the same as having advanced firewalls in that same network. Those small firewalls are not as capable.

Most firewalls sit between two networks and mitigate traffic between them. The main purpose is security of a network, so it filters out some types of communications, like partial packets or specific kinds of attacks that can be used by hackers to get into a network. Mostly, it blocks certain types of communications from entering or exiting.

China's Great Firewall sits on the side and looks for keywords, traps traffic, and blocks content that is objectionable. It blocks Internet addresses, servers or whole domains, like some sexually explicit and religious websites. It has search criteria for news, health, education entertainment, and political issues affecting China, but can filter anything the government may choose. The Great Firewall is like an adjustable information filter put on the Internet.

The Great Cannon is slightly different. It is a black box that can operate as a man-in-the-middle (MITM) attacker; i.e., traffic can be intercepted in route and the content changed or redirected. Man-in-the-middle attacks allow injection of code that can be used to launch denial-of-service or other types of attacks against the target. So, if a website is hosting some religious content that cannot be blocked by the Great Firewall, the Great Cannon can fire off code that will be attached to the sessions of users attempted to access the site. It can modify the content of messages sent to or from the website, block any subsequent traffic, redirect it to a site the government controls, or duplicate it so as to clog up the website of the offending party.

Its use came to light when it was discovered attacking GitHub and Greatfire.org. GitHub is a code-sharing site; it has software that looks for servers that block web connections out of China and was a hosting site for the Chinese edition of the *New York Times*. Greatfire.org is an organization that tries to help people bypass censorship mechanisms, and believes the attack was directed at the software-sharing site rather than others within GitHub.[16] Greatfire.org assumed that the attack was directed by the Great Firewall. Subsequent analysis by security researchers from the University of California at Berkeley, the University of Toronto, and Princeton showed something different, something that complemented the Great Firewall.[17]

The Great Firewall and Great Cannon are mechanisms that can restrict and block, sometimes undermine, communications between individuals. But they don't keep records on individuals that are seen to be on the sending and receiving end of these transactions. That is what the Golden Shield does. As a national database, something few countries are willing to establish, it is fed information from networks that can find, identify, and record information about any individual. Greg Walton describes it this way: "Old style censorship is being replaced with a massive, ubiquitous architecture of surveillance: the Golden Shield. Ultimately the aim is to integrate a gigantic online database with all-encompassing surveillance network—incorporating speech and face recognition, closed-circuit television, smart cards, credit records, and Internet surveillance technologies."[18]

The Golden Shield is a higher level network that does more than information technology, since it has presence on a number of physical sensors that can do facial recognition, fingerprints, credit analysis, and geolocation, combining those with the ability to monitor traffic on the Internet. This is the Big Brother George Orwell taught us to fear. It wields power for its own sake—at least for the sake of the Communist Party. For individuals who chose their own path, the Golden Shield will document the direction they travel and what they do along the way. Those they watch will not come to a good end.

The Golden Shield, Great Firewall, and Great Cannon manage information that enters and leaves China. They cannot control all information, though

China has persistently tried to make them do so. One of the options discussed for retaliation against China's stealing security clearance records from the U.S. was helping skirt the Great Firewall, an effort that would have created tremendous stresses in Party monitoring of its citizens.[19] Even without that formal help, the Chinese people are inventive and have help in skirting the mechanisms that keep them from being broadly informed. That is different from being a dissident. The government can afford to ignore a little of this kind of information gathering, which is not likely to hurt the central government, though it may be more practical than magnanimous.

On the other side of the same coin, China can steal business information and feed it back into the economy, making it difficult to attribute the thefts to China or trace the benefits that come to their economy by stealing. Their control of their Internet works to protect those efforts while keeping their own citizens under constant surveillance. Though the Chinese try to portray their counter-terror legislation as an attempt to deal with terrorists who do operate in China, especially in the northwest, it is more accurately described as a way to deflect the criticism of thefts of proprietary information, especially source code of software made by foreign manufacturers. They can force vendors to supply them source code and encryption software to virtually guarantee that proprietary information cannot be protected inside China.

The legislation seemed to point to terrorism when the first drafts were posted. The issue that mattered most was the turning over of source code and encryption software, but there was concern over something called "data localization," meaning that data collected in China would be hosted in China.

In his almost-defense of the legislation, Zunyou Zhou says the Chinese tried to make their terrorism definition fit an international standard, and bowing to international criticism, dropped the data localization and encryption requirements.[20]

Actually, the Chinese bowed to nobody, and took their collection of software underground because they knew their actions were not going to be accepted in the international community. In May 2016, the reports started to come into U.S. press sources that indicated that the same rules were being applied in secret.[21] Businesses such as Apple were called in to have discussions with representatives of both the military and government intelligence functions about the capabilities and operations of their software. Apple denied turning over any source code, but the long-term effects are not just in software.

The questioning was of concern because it was intrusive and could lead to the disclosure of proprietary information about their products. It could also lead to the disclosure of export-restricted information, which few businesses in China would ever admit to because they would be liable for exporting restricted information and fined for violating U.S. export laws. We will never know all of the different types of information China has collected and used in products it then sells back to customers of its world markets. It is only in the last few years that businesses operating in China have found how disruptive and one-sided these new laws have turned out to be.

What the totality of controls of China's Internet actually does is give it a base from which to hack businesses. More than that, the U.S.–China Economic and Security Review Commission report to Congress for 2016 says it does much more: "China appears to be conducting a campaign of commercial espionage against U.S. companies involving a combination of cyberespionage and human infiltration to systematically penetrate the information systems of U.S. companies to steal their intellectual property, devalue them, and acquire them at dramatically reduced prices."

In terms of cyberwar this sets a new standard for theft of proprietary information and using that information to devalue and buy assets at reduced prices China sometimes creates by having its own state-owned enterprises manipulate business with foreign entities. The best current example is not a U.S. company; it's German.

Before the German government announced it was going to revisit the purchase of a company that it had already approved, not many investors were aware of Aixtron SE. Several business news outlets reported the sale of this company to China's Fujian Grand Chip Investment Fund LP (FGC) in May. *The New York Times* looked into the back story of how this purchase came to be.[22] If we want to play with Chinese investments, it is a good thing to know how they play. More important, however, is the central question the article poses: How do we treat bids that cross between private investment and state-orchestrated takeovers? Does those companies operate with a view towards improving China's position in the global markets, or do they just serve their own business interests?

The cancellation of an order at the last minute put Aixtron's stock on a downward spiral. The company that pulled that order was San'an Optoelectronics, another Chinese company with funding from some of the same people who worked out the acquisition on Aixtron. The story in the *Times* documents the connections between the different companies that were related both to the purchaser and the business relationships Aixtron had in China. This purchase, and one other major one, made Germany the biggest recipient of Chinese capital in Europe. German concerns about technology transfer were not an issue until something else sparked the government's interest.

In November, the *New York Times* reported a new angle on the sale of

Aixtron, the objections of the Committee on Foreign Investment in the United States (CFIUS), a body that usually only gets involved in the sale of U.S. companies to foreign entities. This time, however, CFIUS was interested because Aixtron does quite a bit of business in the United States, though it produces only 20 percent of their revenue. Aixtron makes equipment used in chip making in these categories:

- Compound Semiconductors:
  - MOCVD (metal organic chemical vapor deposition) is one of the most important technologies for producing compound semiconductors, which are an essential element of optoelectronic components. AIXTRON's customers use the MOCVD technology for the manufacturing of different applications such as:
- MOCVD OPTO, a technology for producing optoelectronic devices or LEDs that are widely used in lighting, display application or data communication.
- MOCVD for Power Electronics (PE), a technology used in application areas such as consumer electronics (e.g., in the field of wireless charging), automotive (e.g., components for electrical vehicles and self-driving cars), white goods (e.g., components for more efficient air conditioners) and industrial devices (e.g., components for more efficient wind turbines or high-speed trains).
- MOCVD TFOS (Three Five on Silicon), a technology used for the development of future logic devices.
- Silicon Semiconductors:
  - Atomic layer deposition (ALD) is a process to manufacture ultrathin films for semiconductor components that are necessary for the production of memory chips mainly used in DRAM and NAND Flash devices (e.g., SSDs, USB sticks, memory chips for digital cameras).
- Organic Electronics:
  - Organic Vapor Phase Deposition (OVPD) is a process for the thin-film deposition of organic materials. Plasma Enhanced Chemical Vapor Deposition (PECVD) is a process for thin-film encapsulation of organic layers. These technologies enable the production of organic light emitting diodes (OLEDs), which are increasingly used in displays or OLED TVs.[23]

What CFIUS was investigating was the purchase of technology related to the making of some very sensitive chips used in the next generation of commercial and military applications. It was not disposed to seeing that technology sold to Chinese interests.

The Chinese economy is always thought about in terms of low wages that give manufacturers a cost advantage in making and selling products to their customers, but it is learning to buy those companies rather than work for them.

The U.S. continues to believe the Chinese have few places to put their excess money, a subtle statement that does not reflect reality. It implies that the Chinese will not use their leverage on the ownership of debt. The Chinese have a global economy to put their money into, and they are buying real estate and making corporate acquisitions at an alarming rate. The assumption is the Chinese will not use their position in U.S. currency for their own benefit, which is naïve at best, but at the least, politically motivated. The symbiotic relationship of the U.S. and China is often cited by business leaders as the main reason why it does not benefit China to leverage that debt. On the whole, both arguments are self-serving. To some extent, the Chinese will do whatever is in their national interest without consideration as to how it might affect the United States. The Chinese own $1.3 trillion in U.S. debt, but that is only a third of their currency reserves. That leaves them with considerable money to expand their empire.

While it might be true that a mortgage lender can have little influence over us because they own the major component of our debt, a normal mortgage lender is not a country with a totally different political system that is aggressively competing with us. China can stop buying U.S. debt anytime it wants, and by that action alone, it can raise the cost of future debt and disrupt the U.S. economy. There is probably not a benevolent streak in the Chinese political structure that leads them away from making that threat.

When they cut back on buying U.S. debt in August 2015, they may have been doing so to protect their own currency. Nonetheless, interest rates in the U.S. rose, demonstrating what happens when they stop financing our national diet of overspending. The financial analysts look at China when that happens, but what they really need to look at is the U.S. propensity for spending too much. China owns about seven percent of our debt, even though it is the largest single debt holder. That is declared ownership, not beneficial ownership. We will never know how much of the holdings of offshore accounts like the ones exposed in the "Panama Papers" are actually owned by China's state-owned enterprises. Chinese individuals were certainly involved in some of the businesses identified in the Panama Papers, and have used front companies to gain access to a number of leaders in other countries.

In April 2016, a group of journalists in the International Consortium of Investigative Journalists published an article that named a law firm, Mossack

Fonseca in Panama City, Panama, as the target of a hacker group that stole and distributed some interesting records that showed the creation of holding companies that protected the real owner from detection by prying governments who might want taxes or state riches returned. The Infosec Institute published an analysis of how the hacking occurred and what was stolen.[24] The article was developed from work done by the same hacker who identified vulnerabilities in websites at the *Los Angeles Times, New York Times*, NASA, and Edward Snowden's own website. He points out that the systems at Mossack Fonseca had enough known vulnerabilities that the information could have been taken by a knowledgeable hacker from almost anywhere. The e-mail server was hacked and over four and a half million e-mails were taken. Three million database files, two million documents formatted in PDF, and a million images were included. There were a relatively small number (320,000) of other documents taken. What was taken pointed to some prominent individuals including Vladimir Putin and Iceland's prime minister, David Gunnlagsson, who resigned after the disclosure. Some of the others were associated with people in China.

Documents leaked from Panama name family members of the Chinese president, Xi Jinping, and two other members of China's Standing Committee, Zhang Gaoli and Liu Yunshan.[25] But the documents also include billionaire Li Ka Shing; Thomas and Raymond Kwok, whose Hong Kong property empire is valued at $14.7 billion; Hui Ka Yan, who had been a member of the National Committee of the Chinese People's Political Consultative Conferences (CPPCC) from 2008 to 2013 and thus was a "Politically Exposed Person, which required Enhanced Due Diligence"; and Chinese billionaire Liang Guangwei, a former People's Liberation Army soldier and head of a state-backed technology conglomerate who recently bought a $64 million block of land near the headquarters of an Australian spy agency.[26] The cited sources leave little doubt that Mossack Fonseca knew the political sensitivity of some of their clients.

Panama is not the host for the largest number of companies that are sometimes called shell companies, but with 350,000 it ranks third behind Hong Kong and the British Virgin Islands.[27] If each of the other two had only the same amount of shell companies it created, there are over a million companies like the ones discovered in Panama. This one company was not unique in the world of law firms, where many churn out tax havens, trusts, and corporations that protect the legitimate business interests of clients. In most cases, the law firms either do not know who they are protecting or representing, or do not want to know, even finding ways to not identify them.

Chinese company executives are given jobs because of their positions in the Party, as much as because of their business acumen, but these papers show they have business knowledge too. Most removals from businesses positions had to do with a business leader taking a track that was not in line with the central government leadership. So we might think that this action by ZTE would represent a set of "rogue executives" doing something the government did not condone or sanction. Usually, that kind of action results in a business leader disappearing and not coming back, while he is questioned by the government. That did not happen in this case.

Clare Baldwin, writing for Reuters, says the ZTE representative, on a 3 April 2016 call, claims these management realignments take place every three years and this was just one of those events. However, he also said, "ZTE spokesman Dai said he could not confirm which executives would be involved in the management changes to be announced on Tuesday, and could not comment on whether the upcoming changes were related to the alleged Iran sanctions breach in any way. 'I cannot speculate on this type of discussion,' he said. 'I am not in a position to comment.'" Treating the move as a normal action indicates that the board members were not involved in the kind of activity that the government objected to.

An *Asian Age* story on the same thing treats this like a spat that has nothing to do with ZTE.[28] The U.S. sanctions caused the individuals to be removed. The article does not mention that the ZTE internal documents cite the setting up of dummy corporations and defined export rules that would be violated in doing so. They also clearly show that other Chinese companies were doing the same thing. No company in China is going to sell anything to Iran without the central government knowing about it and sanctioning it.

At least we now know what was agreed to settle the clear violations of U.S. Export laws by ZTE. What we don't know is why that kind of action was acceptable to the U.S. government. Treating this as the action of a few rogue executives ignores the role of the central government in controlling Chinese companies. Investigators have recently called for documents from Huawei, Inc., as a continuation of the ZTE exposure. ZTE's internal documents mentioned another company that was exporting technology to Iran, and others, and the investigation centers around whether Huawei was that company. The world business community seems fearful of this kind of investigation because the inference is that any company might engage in this kind of behavior.

U.S. businesses still line up to sell to China, a lucrative market. At the same time, China makes more of the world's goods and services than at any time in recent memory, overtaking and passing the United States.[29] Both sides of business get what they want from this arrangement, but we have to wonder if the citizens of their countries know what they are doing. If China were our enemy—if we were at war, and we recognized it as war—manufacturing would certainly change, though not likely stop. Even in World War II, Ford had a plant in Nazi Germany that continued its operations; banks in Switzerland continued their money transfers; and airlines, trains and ground transport continued to move between German cities. Yet at the end of that war, almost nothing was written about the collaboration of U.S. businesses with that enemy. The records of companies still doing business in the U.S. today are bottled up in company offices or buried with the dead.[30]

When it comes to networks, most companies are international, sometimes without knowing it. It is difficult to think of a large company that has no overseas operations of some sort, and companies such as IBM, HP, Siemens, McDonald's, China Mobile, Oracle, GM and BAE have thousands of employees in other countries who are directly or indirectly contracted.

All of these have major computer systems and networks that interconnect the world and give them connectivity to their mother ships through those overseas circuits that transport corporate networks. They connect to major business partners, customers, suppliers and their own business units. These change over time; sometimes they grow as they acquire one another; sometimes they go bankrupt and sell off all the assets. Most companies don't like to use the Internet for this kind of thing. They like to separate themselves from it if they can, mainly to protect their internal communications, but they used to be better at it than they are now. We can blame the economy for that, but it would be only partly true.

Businesses are being pressed to find more efficient ways to work because they have competition from places with lower labor rates and cheap currencies, such as China. So they cut staff and try to find cheaper ways to do things. Corporate IT staffs are shrinking and so are their security elements, and there is no way to do all that is needed. One blogger said he had bought security equipment to install, but they were so short of staff that nobody did install it. They are reducing hardware costs by supporting computers that do not even belong to them, allowing people more leeway in working from home, on their own computers, and connecting a variety of smart phones that can expose their business to greater risks. They shift to wireless office lines because that is also cheaper and can be moved when the lease runs out. And they outsource to other companies what they can. These actions all raise risks because they are moving the responsibility for protecting corporate secrets away from the security staff to a user or a partner. At the same time, hackers are targeting more businesses, having more success, and not having to work nearly as hard to do it.

What is hard for most people to see is when this becomes a serious issue to the U.S., as a country. A few hackers hit businesses every day and some of them will swindle some ladies in Arkansas, but that doesn't add up to a national concern. That is because most of us never get to see what the government sees at the top. It gets reports from the CIA on what is going on in the world, reports from National Security Agency on what is happening in the networks of the government, and reports from the business leaders of the major companies in the USA about what is happening to them. This is not security of some data in a computer network. This is national security.

All countries have laws about national security. Just to be clear, they are only worried about their own, not ours. They tell other countries how they are allowed to transmit things through their countries. They spy on the rest of the world to get more information. They share things with other countries that think like they do and want to share. This is usually considered to be "legitimate self-defense" or some such thing. There are some grey areas here, of course.

Economic warfare is just one of them. Can we steal the bids from the new ship that China wants to build and give them to an ally of ours so they can build it? Can we tell GM and Ford what we know about Chery, the Chinese company that makes automobiles? If a well-meaning person gives us the plans for that new bridge in Kalamazoo, can we let our bridge builders see them? In our country the answer is always no. We don't even let our own companies bribe officials in other countries who will not give bids to people who won't bribe them. We could do all of these things, because it is in our power to do it, but it would violate some law somewhere that was written before we became a growing economy in the larger world.

Every country requires access to our computer networks to monitor traffic passing through them, mostly e-mail and data fields. They say they will not keep this stuff unless it points to a crime or terrorist activity. They read our mail; they listen to phone conversations. They record a lot they can't listen to, and this stuff is stored all over the world in various systems. I remember when East Germany fell and the press started to pour through the Stasi networks of surveillance. People were surprised by how detailed it became and how much information could be collected on almost anyone, if they had the right equipment and the will. There were logs and records on daily conversations, and videos to go with them.

Surveillance and monitoring are becoming a science, and most countries are good at it. That is what national security is all about. When the Middle East started to come unglued and there was the collapse of regimes there, the amount and type of surveillance was one of the first things to come to light. The second was the number and types of companies that were making the little boxes that did it. Bruce Schneier had a blog article on some equipment from England that could block cell phones in a particular location, intercept them, or get them to transmit codes that were unique to each one that would allow the calls to be traced back to the owner. Timothy Karr of the Save the Internet Foundation pointed out that equipment is used in Egypt to monitor names and addresses of people On Facebook, Twitter, and YouTube so people could be watched with greater accuracy. The company that gave them the equipment, he said, was Narus in Sunnyvale, California, formed by some former Israeli intelligence folks. That company belongs to Boeing. A Finnish newspaper, *Helsingin Sanomat*, reported that Nokia had sold a "spy network" to Iran that could monitor voice and data, pick out target information and flag it. It can monitor voice, data, instant messages, mobile phones and fixed landlines, e-mail and fax. Nokia says it was a "test system" that could not be used for the fixed Internet. We could only wonder why they would sell them a test system to begin with. What were they supposed to be testing? This is the technology needed in the Great Firewall.

It sounds like a good thing to help out law enforcement, and that kind of logic is what allows the export of these types of things to other countries. What is missing from these license applications to a government is the nature of the crime where the equipment will be used. Speaking out against the Thai emperor can be a crime and 2,000 websites were on their blacklist for doing just that. Complaining about the solution the local government has offered to fix that dam may be a crime. The term *honor crime* is certainly not a term that has anything to do with honor. In some countries, if my daughter decides to marry someone who is not of the same religious sect, she might be killed. The people who kill her might be investigated but not arrested. Rape is somehow excused in these kinds of cases and the victim blamed for it. One of the news shows carried an interview with a woman in Afghanistan who had had all of these things happen to her at once and they were going to hang her for the finale. That is not easy to understand.

My son sent me a picture of an airport security checkpoint with a sign that said, "Possession of drugs is Punishable by Death." Americans passing through can certainly see the law is different there. In a stay in Greenland, I got to see how Danes handle drug enforcement. We had a guy come up to our military installation, from Copenhagen, to play the piano. He was a heroin user, which we consider a crime (carrying the things needed to take illegal drugs and possessing them), but we were not under U.S. law there. A storm put him in the position of being out of drugs and no place to go. He turned himself in and asked for methadone. He could have gotten it from the Danes, but the only hospital for 300 miles was the U.S. military hospital, which didn't carry it. Because he was ill, they put him on the next plane, which was going to the U.S., not Copenhagen, and, on landing, he was arrested for transporting drug paraphernalia. It was justice, but it was hard to tell what kind. The definition of "crime" is different everywhere.

The Chinese like to say that the business information of a government-owned business is national security information—a state secret. They leave that definition vague and let companies figure it out for themselves. The laws covering the kinds of businesses that can be purchased by foreign entities are equally vague. If a business decides that it wants to buy goods from one company but want to find out what the competing prices might be, it will be hard to do. If it decides to buy into that same company, it is a lottery of interpretation and not a straight business decision. The Chinese are not alone in this kind of thing, but they are at the top of the target list for people who are looking for abuses of it. Since computers save everything, it is not easy to know that a business has some of the things China would call state secrets.

Every country has an array of equipment that allows them to monitor other countries' people, on the off chance that they may be violating the law. They call this national security monitoring. If you are texting, IM-ing, e-mailing, or are Facebook friends with someone from another country, it is likely you are being monitored somewhere, by somebody, and probably more than just one somebody. This is not spying and it is not illegal in any country that does it. You could be committing a crime you don't know about, or have ever had described to you. Ask Liao Yiwu, who spent four years in Chinese prisons after he wrote a poem called "Massacre" following the 1989 demonstrations at Tiananmen Square. Writing it takes more courage when you know you might be living in prison afterwards.

Moviemakers can show how it is possible to get on a computer and hack almost anything, anywhere, and change information or records, shut down electric grids and open dams to let the flood waters out. Several movies and television shows popularize the myth. If it were that easy, we wouldn't have computers. They would be banned or controlled because they are too dangerous. Our bank accounts would not be safe. Our personal information would be public knowledge. We couldn't use credit cards at all, although I wonder how we do that now, what with all the theft of their numbers. They might even be controlled like guns, registered and have some limits put on who is allowed to have them. Hackers are smart enough to not want that to happen, and so are governments.

I asked a hacker who demonstrated his skills at getting into some of the most secure computers we had why his friends in the hacking world had not brought down most of the Internet. They had the ability to come close to doing that, and we always thought they stopped short. He said, "Because they use it." He was pointing out that bringing it down would have long-term consequences that none of them wanted to live with and would make their job harder. There are more than a few red lines crossed. It would deprive them of some of their best targets and improve the defenses of the rest. Better to leave well enough alone. Better to not make war.

Although we have had some fairly spectacular hacks in the past year, at Yahoo, the Democratic National Committee, the Office of Personnel Management, Sony, Lockheed Martin, and others, the business community depends on the trust we have in their ability to make the Internet safe for commerce. If we start to think it is not safe, Baidu, Alibaba, eBay, Amazon, and Google will not be in business much longer. Hackers are starting to push that limit. If someone wants to undermine the world's economy, that would be a good place to start, but it is more difficult than those movies and TV shows would indicate, and may have the type of unforeseen consequences that hackers know about. Nobody, including the Chinese, wants to kill the golden goose, even though it is on life support at the moment. What put the Internet at risk happened quite some time ago.

Starting in November of 2010, several systems were hacked by someone who established over 300 control systems, almost all around Beijing. What made this different from other attacks was that the attackers were going after a place called RSA that was famous for its ability to do encryption of various sorts. RSA makes a token that verifies authorized users through a home network. You would think a place that makes security devices would be secure.

During the next few months, several other major companies were hacked in the same way, and there was a pattern to these that will make anyone who sees the list nervous.[31] There was the IRS; USAA, which primarily handles insurance and banking for military people; several locations of COMCAST and Computer Sciences Corporation; a few locations of IBM; the U.S. Cert, which handles investigations into computer incidents at the federal level; the Defense Department Network Information Center; Facebook; Fannie May; Freddie Mac; Kaiser Foundation Health Care System; McAfee, Inc., the antivirus people; Motorola; Wells Fargo Bank (and Wachovia, now owned by Wells Fargo); MIT; University of Nebraska, Lincoln; University of Pittsburg; VMWare; the World Bank; and almost every telecommunications company of any size, anywhere in the world. That last one included all the major telecoms in China—so they are hacking their own telecoms. It is almost like someone said, "Go out and get everything you can."

There were 760 companies in all, and 20 percent of the Fortune 100. This is the kind of attack, spread over several months and extremely successful, that can get our leaders excited and ready to do something. One of the companies wanted permission from the feds to go after the people that were behind this and find out where and who they were. That would be nice to know.

There is a breaking point in a relationship with another country that can come without warning or understanding of how we got there, and this instance was close enough for most people. It is time to do something so the ones who are doing this understand that we cannot have folks hacking into the foundations of the culture we have here. The only thing missing on the list of 760 was churches. It certainly looks like the Chinese did it and, if it wasn't the Chinese doing this, then they need to find out who did, and stop them from operating in their country. It is too much like war for comfort.

What the Chinese have seemingly done is combine their military, academic, and criminal organizations into a more capable bunch of hackers.[32] Then they complain that they are unable to control this type of hacking by criminal elements, just as the U.S. is unable to control hacking that comes from us. It is just another way of establishing deniability. It is a deception.

China has much more control over what their people do and do not do than the U.S. does—probably more than any country in the world. When one of their gangs gets caught running drugs or stealing a truck, the guilty don't go to jail for a few years. They get a bullet in the back of the head. In the 1990s, China executed more people than the rest of world's countries combined. During the anti-corruption campaigns of the military-managed businesses, 16 people were executed, to help make the point. Everyone involved got the message that China's tolerance for aberrant behavior is much lower than the rest of the world.

The Chinese can protest all they want about their inability to control their criminal gangs, but they would be hard-pressed to say they can't control their activities on the Internet. They control everything on the Internet, and we don't have to look much further than the Golden Shield to prove it. Their view of the Internet is the opposite of ours.

Intelligence Recon

There is nothing wrong with countries spying on each other, but there are rules associated with it. If you see those CIA folks on television, you would think they were allowed to do most anything, including killing members of our own government to hide the secret society that really runs it. They really can't do those kinds of things without getting in a lot of trouble. Those who don't know the rules are not spies or have never been involved in spying.

Government spying is very complicated and involves quite a few people. Our government says we spend $80 billion a year doing it, but we can only guess how many there are, since that kind of thing is classified by the government. We won't see the Chinese intelligence budget in print anywhere, but we will see the results. That is probably a better way to judge whether something is working anyway.

If I decide to spy on someone in my neighborhood, there are a number of things I can do. I can research his housing plot and the house itself, the cars that he owns, his trash to see where he works and what he buys. I could mount a camera on my roof and look over his way. This is called "open source" collection because some of the information is not protected by anyone. I might not tell people why I was doing it, or may want to avoid going through his trash while he is still home, so I have to be a little careful about how I go about it. It is really only partially "open" in that sense. I always think about this when I hear that our number one export to China is trash, and 80 percent of discarded computers and cell phones end up there. We do a lot of shredding at our house.

I can also follow him, talk to his neighbors, his co-workers, his friends at church, or videotape him at various places, though this is not in my nature. This is *stalking* in some circles, so it might also be against the law. I can get a job where he works, or get a job working for him. This is useful if I can stay there awhile and get to know this person. He sees me every day and it is easier to find things out. I can become his friend and visit his family and friends. I can download company information that he has and read things he writes. There is quite a bit that I can find out without going any further in my methods.

If I don't mind doing things that might be criminal to find out more, I can listen to his wireless calls, intercept his e-mail, open his mail or packages, use imaging equipment to observe heat signatures in his house, and plant electronic bugs in his cars, his work area and his home. If you are saying, "That isn't very nice," you have discovered the essence of spying. It isn't. Rupert Murdoch can vouch for how people feel about it.

Spying is the collection and analysis of other countries' secrets, and it is pretty sophisticated compared to what I can do with my neighbor. It is divided into a number of things called "INTS" that are general categories of capabilities to do things; we have a range of other intelligence collection capabilities called COMINT, SIGINT, HUMINT, ELINT, MASINT and the like that describe a certain type of thing being collected and analyzed. HUMINT, for example, focuses on information that is collected from human sources. Interviewing my neighbors about the house I was interested in would be one form of it. COMINT looks at his communications like his cell phone or computer. Rather than try to figure out all of the different types, it is better just to think about them as spying, using different technologies. These are just ways that governments collect raw intelligence. They have a lot more money than people like us.

If I could put all of the government's spying capabilities in my hands for a few days, I might not be able to review all that can be collected before I died. I can have enough information on this guy to write ten books, and then some. Governments have a lot of resources the rest of us don't have, and they have rules about how they go about collecting and using information, including sharing it with me. That is not going to happen. Just to be clear, I'm not spying on any of my neighbors either.

The Chinese are good at human intelligence, placing people in a country to work and having them check in every now and again, but not spying on anyone until they are needed. Let's take the case of Dongfan "Greg" Chung, a naturalized U.S. citizen who had been in the country for 40 years. He got caught spying for China because another engineer, working for a different company, was caught, and the FBI started watching his Chinese handler. "Greg" worked for Boeing and had some Boeing documents in his trash. He had 300,000 documents in his house when he was arrested. His prosecution was under a new 1999 law that made it a crime to steal trade secrets.

It just shows that spying is against the law most of the time, but not all the time. It depends on what country it is, why it is a crime, and how stupid it was. It is hard to keep a really stupid thing, or a case that has to go to court, out of the press. You hear every once in a while that so-and-so tried to sell documents to someone in the FBI and you wonder how they could be so dumb. Well, the FBI is not going around saying they are the FBI at times like that, and they don't dress in a suit and tie when they visit this type of person. They dress like the person expects a criminal to look, probably like any of us on a Saturday morning.

Real spying is not war and is done by almost every government in the world. If the Chinese spy on us, they do it with the clear understanding that we spy on them too. Every country spies on the others as much as they can support. In all the world's governments we understand spying and expect it. That won't help you if you are caught spying in the U.S., because it is criminal. Even if something isn't an act of war, it may still be criminal. It is just one of those little quirks in the way the law works. We don't go to war because someone commits a crime in our country, or when a country spies on us, or we would be at war with every country except Greenland.

National Security and Business Spying

In the past few years, there has been quite a bit of testimony at hearings on Capitol Hill about China spying on U.S. business. Part of the result was a new law, the Defend Trade Secrets Act of 2016. That law allows the federal government to have a course of action against an entity that steals trade secrets, but it has adopted the short statute of limitations of three years of the Uniform Trade Secrets Act. It takes longer than three years to discover an electronic trade secret theft, acquaint lawyers with the information being stolen, and bring an action.

The Chinese have developed an approach to stealing technologies and secrets that is thoughtful and takes a long view. They don't get in a hurry. In the days before the Internet, I interviewed a scientist who traveled extensively to Russia and China because he had an expertise that very few people had. He asked to see me because I was briefing people like him who had high-level security clearances and traveled to countries where people might want that information. He thought these briefings were effective but did not achieve the right result. He put it this way: "Dennis, if you tell people there is a Russian under every bush, when they get to a place where there are bushes, they will want to look under them. The secret police will see them looking under the bushes and think that is suspicious, so they will follow them. You tell them they might be followed, so they are looking for that to happen, and that makes them look more suspicious. Let me tell you the differences between how information is taken from people in those places, from my own experience." He then went on for over an hour.

The Russian approaches to him were crude and he usually could see them coming. His interpreter would ask him questions that seemed to be from outer space and totally unrelated to the topic of conversation. He was offered a prostitute and a hotel room to take her to, but he declined. He said he was too old to enjoy such a lovely person. He met strangers at the hotel who asked him questions about his reasons for being in Moscow and chatted about his particular technical specialty as if they had been doing it their entire lives. Some of his fellow scientists had large-bodied men following them around, and they didn't seem to know very much about the conference topics. They were there mostly to prevent defections, which went on anyway, but less often when these guys were around. Their presence discouraged much dialog, but at the same time, since they didn't know the subject at all, the scientists could talk about things of mutual interest. Both the scientists and the KGB knew the game and how to play their part in it.

On the other hand, he attended conferences with his Russian contemporaries; they would be talking about some obscure area of this technology and would reach a point where state secrets were involved. He could tell them the topic was a state secret and they would veer away from it. They recognized that some secrets would not be shared between them, but they could still work together.

The Chinese were different. They also supplied him with an interpreter, over a period of several years, the same one. The two of them had an understanding that she was asked to take care of whatever needs he might have. He was pretty sure that included sexual needs, because she had mentioned the possibility—just once. He said she would sometimes get close to him and he felt like there was an attraction between them, but she never made a move out of turn. They had a good relationship and it was businesslike, but friendly. That was the way he wanted it to be.

She took him to several universities and conferences where he was asked to speak. At one university, he was asked a question about an area of his work that was a state secret. He said he could not reply. The student almost jumped out of his chair to respond, "Why can you not tell us now? A hundred years from now it will not matter to any of us and we will both have the same information you keep from us now." He said he actually thought about this for along time after, because it was true. He said he could not give out that kind of information and there was murmuring from the audience. He stopped answering questions and left. He never heard that question again from another audience, but he did find out how they were able, years later, to get the information he was protecting.

An associate professor at one of the universities in California where the scientist taught was invited to speak at a conference in Beijing. He was not asked very often and was glad to go. Over 100 people met him at the airport. There were various academics, politicians, and people to handle every detail of his trip. He was invited to a dinner that evening and almost the same number of people attended, toasting American friendship. He was the only American there.

At the conference the next day they asked him all kinds of questions and he did his very best to answer them all. He "talked his head off," was the way he summarized the outcome. They were willing to spend time and resources to find the right people to get the information they needed.

A Buildup to War

Where it relates to national security, China has a number of collection programs that have been successful focusing on classified things like weapons and government programs. This is quite a bit different from stealing things from networked computers, because the people who have this information have better security. They encrypt most everything and compartmentalize the information so not everyone has access to everything. The networks are very restricted. About the only way to get access to secrets like that is to have someone who works at a place with the information steal it. That is the hardest and riskiest spying there is.

The Chinese are not the only ones who steal information from us, but the people getting caught at it are, increasingly, linked to China and not Russia. One of my professors used to remind us that criminals in jails are not a good measure of the capability of the criminal community at large. Government-sponsored attackers will be careful and good at what they do. For every one of those people who get caught, there are a few really good ones who rarely do, especially ones who have government protection. They are good, and getting better.

The Chinese have been doing the old-fashioned kind of spying, and that is also increasing at the same pace. The year 1985 was the "Year of the Spy." It was a big year for industrial security because there were so many examples we could tell our students about. One after another, they were identified, went to court, splashed the terrible things they were doing all over the news-papers and magazines, and went to jail. We thought it was wonderful.

Twelve people were prosecuted that year, several more in the years before. I was teaching a course in industrial security in Palo Alto, California, and came downstairs to see press in the lobby of our hotel. A secretary to the president of Systems Control, Inc., down the street from where we were, had just been arrested for her part in helping a fellow named James Harper, her husband, steal classified information and sell it to Polish intelligence. We didn't get much press coverage of our courses, so we were all a little startled to see cameras and the director of industrial security for the region setting up to make a public statement. There was an entire row of reporters sitting in the back of the room. This was not even close to what we have had recently, yet very few people are paying attention. There were, since 2008, 57 defen-dants in different courts charged with spying for China.[33] In the annuals of spying, 57 federal prosecutions in a 3-year span is a pandemic. It is so many that it is a little hard to believe that more people haven't noticed.

In 2016, Admiral Rogers, director of the NSA, testified before the Senate Armed Services Committee. He said the Chinese have not stopped spying on U.S. industries in spite of an agreement to do so, but we are no longer sure if they are putting stolen business secrets back into Chinese companies.[34] That does not seem like a very satisfying answer, given that he would likely know the complete answer to that question.

FireEye's Laura Galante put it a slightly different way in an interview with *Fortune*.[35] In many cases they observed, the Chinese are getting into the networks they may need information from, but they are not taking data from them. They could, if they so desire, but they do not. The number of attacks have gone down, but they are persistent and focused. That is probably not good news to those looking for a slowdown in Chinese hacking.

This kind of activity makes the theft of data harder to detect. Security people look for exfiltration of data, and attacks that allowed a person into the network. Combine this with the relatively new movement of Chinese hacking out of the army and into more secure and tech-savvy agencies, and we should not be happy that the attacks have slowed down when all that does is make them harder to detect.

It would be easy to say that everyone spies on each other, but it would not be right to say that the Chinese aren't doing more than their fair share of it. For one thing, for every case we prosecute, there are a few more that are going on that haven't been found. What we are seeing is just a small part of what is probably going on.

In the past 15 years, China has stolen classified details of every major nuclear and neutron bomb the U.S. had in its inventory.[36] They have had ongoing espionage activity at the nuclear laboratories (Los Alamos, Lawrence Livermore, Oak Ridge, and Sandia) that produce and develop the weapons. This allows them to make their weapons smaller and easier to shoot a long way on a missile. According to a House Select Committee report,

> The United States did not become fully aware of the magnitude of the counter-intelligence problem at the Department of Energy national weapons laboratories until 1995. In 1995 the United States received a classified PRC document that demonstrated that the PRC had obtained U.S. design information on the W-88 warhead and technical information concerning approximately half a dozen other U.S. thermonuclear warheads and associated reentry vehicles.[37]

Among secrets, nuclear weapons secrets are some of the most valuable, and closely guarded, we have. When the Chinese have them, it doesn't speak well for our ability to protect anything from them. If they have those, they have a good deal more too.

China has stolen U.S. missile guidance technology and exported it to other countries like Iran, Pakistan, Syria, Libya and North Korea. It sold medium range missiles to Saudi Arabia and trades extensively with Iran,[38] which is not our best friend after trying to get Mexican drug gangs to hit embassies in Washington, D.C.—although they did say they didn't do that.

There is a certain amount of risk in any of these types of thefts, but there are ways to reduce that by doing the spying inside in a U.S. business. The other day, I stumbled on a company called Verizon with the ".cn" after it and went to their website to see what Verizon had over there. That site latched hold of my computer and wouldn't let me do anything until I allowed access to my systems. I wouldn't, and had to reformat my hard drive to get them out of my Mac. I didn't know there was a Verizon connected to the China Internet domain, and sent them a note about what happened with their company.

I checked AT&T and found that they had been operating in China for 25 years. Deutsche Telecom is there and is forming a partnership with Huawei to build a cloud infrastructure. There is a Ford China, a Sony China, an HP China, a Starbucks and a list of 250 others that still does not include all of them. We have connections to networks for almost all of our major companies that operate in China, and they work two ways. Those employees are employees of Ford and they are in the networks of Ford, having access to what most Ford employees do. There are restrictions on what a person can see in all corporate networks, but it is an inside connection that starts all of this, and those are much harder to control than the external ones. They don't have to spy to get information through these channels. These are legitimate businesses.

The countries that do spy on businesses, and share that with their own companies, could go out and collect things off the Internet and nobody would mind that. There is even an acceptable range of things that businesses do to spy on each other—what is called business intelligence.

Associations, conferences and trade shows are good places to meet people from the competition. They get to know people and exchange information about the companies who are trying to sell products. There is nothing wrong with this kind of thing, and it is expected that any person traveling to a show will bring back any information about a competitor that might come their way. There are usually booths, shows and meetings in hospitality suites where the conversation is always worthwhile and the food and alcohol are free. It is acceptable to send slide briefings and promotional material to other folks who ask to help promote business. A lot of information is left lying around on tables and anyone can have it. We were teaching a course in industrial security in San Francisco when a gentleman "from Taiwan" came into the back of the room and started collecting our course handouts. One of the other instructors stopped him and asked him where he was from, and he produced a business card from a shipping company. We took the things back from him and had the hotel security showed him out. He may have just wandered in, but it seemed like he knew what he was doing there; else why would a guy from a shipping company be interested in U.S. industrial security?

From the contacts made at these functions, some things can develop. We can set up a request for information from some of them, a greater level of detail that is needed to decide whether a product is directly suited to a type of business being looked at. The federal government also issues these if a business takes the time to register and get the notifications. They reply to these and get into conferences where "potential bidders" are invited. Between businesses and government conferences, a person can stay pretty busy and collect quite a bit of information on what these various organizations are thinking about for the future.

After this round, we could set up additional meetings with targeted companies and visit the sites where the merchandise is made. Before we get to visit another facility, we are likely to have to sign a non-disclosure agreement. It says we don't disclose any of the secrets we will see at the place to anyone else, even to other people in our company who did not attend. These are usually called "site visits" but they are really just specific sales presentations, focusing on a product a client seems to be interested in. So the "don't tell anyone else" clause is not taken very seriously. Smaller companies don't ask very many questions of someone requesting this kind of meeting, but the bigger ones want to know that the company is big enough to justify the time and money spent on putting these together. After these are over, we can arrange for technical interchange meetings where very specific technical topics get ironed out. By the time the whole process is over, a good deal of information will be doled out, but it is generally not going to be spying, by any definition. It is more like fact finding.

Now put this in the perspective of a front company or a business that has a role in spying for a particular country, and you can see how it creates the ability to collect useful things like the names of employees in specific technical specialties, the business structures and where all the offices are, where specific things are made, something about the capitalization of the company, and how they seem to be doing in their business. Once we have a little more information about them, they can be invited on official visits to our company in China.

We can set up a joint venture or a trade agreement with various technical companies that we need to buy goods from. We can reverse engineer most any product we get and figure out how it was made and the materials that would be required. We can then cost those materials and compare our costs against theirs. We probably have not violated the laws of either one of our countries while we were doing all of these things, at least to the point of being prosecuted for it.

The Chinese are doing more than that, by following an information warfare strategy that is much broader and deeper than just the usual collecting of things. They are stealing from businessmen and computers that are supposed to be protected.

They steal quite a lot of proprietary things directly from the contractors, but don't confuse this with cyberwar. It is state-sponsored stealing, which is different. The popular press has confused the two, though, and it is probably understandable since it is sometimes hard to say where one starts and the other leaves off.

If I steal information about the internal computer switches at Comcast, I might be doing it to prepare for war; I might have the idea to steal service from them, or both. I might just want their software, which is theft. It really depends on intent—how they expect to use what the attack has provided. It could be collection of information for some intelligence purpose, a criminal intent, or it can be used in some aspect of war. The information is the same. If I collect it, I try to save as many valuable things as I can, without really knowing what might be useful in the future. Everyone ends up keeping a lot they don't ever use.

Businesses are pretty smart about these secret things, and protecting themselves against people stealing information is something they try to do. This is sometimes more complicated than a person might think. There are a few things working against success. National security policies are the first of them.

Research in Motion (RIM) bumped up against the United Arab Emirates (UAE) and India over how encryption is used to protect commercial e-mail. Blackberry networks are encrypted, so there is nothing new about protecting business interests with encrypted software. But there is always a hidden trade-off in making systems very secure. What RIM does to secure their servers is a series of mechanisms to make its business e-mail reasonably secure against interception or tampering. Most of the time this is a good thing, until we have a terrorist or drug dealers using the security features to protect their business interests. The ability to monitor them is a matter of national security.

If the Chinese do not like the encryption Apple does on the iPhone, they can ask them to provide the government with a decryption key, or they can go to the company using Apple equipment and ask for the data from them. A key, or the data decryption of the e-mail, can be justified on national security grounds. Vendors have stopped keeping keys to the systems their customers use, taking them out of the middle.

It is important for the intelligence communities of the world to be able to get information about what other countries and groups are doing, so Apple is not alone. There are national interests at stake and terrorists to deal with, but what cell phone companies are doing is securing business systems. They are trying to make systems secure so businesses can protect trade secrets and operations from thieves, extortionists and others on the Internet trying to make a profit by access to private e-mail. This is a very fast and growing area of business in its own right. People steal information for profit. We certainly could design systems that would be so secure that nobody could get into them. But there always has to be a balance between the national interest (finding terrorists is just one example) and our business interests (protecting e-mail).

It gets more complicated because the national interests of one country are not always the national interests of everyone else. For some hackers, there is a job opportunity in there. Those in the oil business see all the tricks that are being pulled and they want to know how to stop it, or get in on it, depending on which side they happen to favor. They want to get the information. Businesses try to stop them.

If they all banded together to build entirely bullet-proof computer systems, no hackers could get in for a little while. It would not take them long to discover what these systems were doing to make themselves more secure, and to start looking for ways around it. So they start figuring out what they have to do to get in. At the same time, they could use the techniques that were making those systems more difficult to get into, to improve their own security. At a country level, it would be an arms race, of sorts. The intelligence services don't like this kind of thing, because it makes their work harder too, and they have a bigger stick. The equipment vendors will usually lose out to national security.

When the U.S. complained about China's hacking into our systems, China said more hacking comes from the United States than from China to us, and we should stop complaining about what they are doing. The Chinese were right, though it is close. They have had days when China had many more than the U.S., but some days it is the other way around. On those, more attacks were coming from the U.S., but nobody is saying where they originated. China was implying that the systems were in balance and that the hacking was equally spread across the world; we need to leave them alone. Attribution is not good enough to prove things one way or another, and they know it.

Hackers try to bounce around from one country to another to prevent someone from figuring out where they actually are. This is harder to do than it sounds because it is work to create accounts that are difficult to trace to their original owner. A series of those are needed to make "hops" that can't be traced to their origin. They would rather have people in China believe they were coming from the U.S., which has more computers they can use to target other systems, than using someplace in Estonia, Greenland or Iceland, where they would be easier to find. It is a complicated game, but we should still be able to find and stop them.

Where they tend to overlap is in the defense part of the intelligence community, the Defense Intelligence Agency, the National Security Agency, the National Geospatial Intelligence Agency, and the intelligence parts of the military services. These latter agencies are conflicted because they are both military (sometimes referred to as Title 10) and IC (Title 50) agencies. They tend to follow the rules they like from defense or intelligence, and play both of them against one another. This is mostly in fun, of course, but it is entertaining to watch.

Each federal agency then makes its own policies from the policies of the Director of National Intelligence, Defense (through the Committee for National Security Systems), and the National Institute of Standards and Technology. It is not as confusing as it sounds, and it is one of the major reasons the government gives businesses less protection than they need. Businesses are supposed to be able to cooperate at the national level through the critical infrastructure apparatus run by Homeland Security and the White House. During the years I sat on the Industry Committee of the president's critical infrastructure protection committee, we had difficulty getting agreement among ourselves, but the underlying issue was always liability, something the industry fears. Nobody was willing to step up to collective security requirements or agreement on what could be done to strengthen defenses, even though financial losses were climbing. There was a general fear that liability concerns outweighed actual financial losses on the Internet. Knowing what the vulnerabilities are, and which company creates them, can potentially lead to liability.

As an example, the world had a worm that propagated code for computers; the way it worked was by encrypting those computers and issuing a ransom note to their users. This affected Microsoft computers, but those computers used an operating system that was no longer supported by Microsoft. When the Dark Web started to sell software that exploited a known vulnerability in that software, Microsoft could have said, "We no longer support that." Legally, they might have been on good ground with that, but they knew that the court of public opinion is not a real court. Instead, they issued a patch for the vulnerability.

For two months before the worm struck and the ramsomware executed, the patch was out there waiting for anyone using the old operating system.

Those who did not install the patch were called victims by some, or defendants by lawyers who attempt to show how negligent they were in not installing patches for known vulnerabilities, causing their hospitalized patient clients great harm. That is only just starting to play out.

Outside of government, where parts of critical infrastructure protection are done, each legal business entity has its own networks that are corporate assets, some almost as big as the federal enterprise. Each manufacturing facility, pharmaceutical company, railroad, telecommunications company, service provider and software manufacturer has its own. The national enterprise is made up of the federal enterprise and the legal business-owned networks that operate under U.S. law.

At the level just below the National Security Council, the president's Critical Infrastructure Protection Board (CIPB) was supposed to be developing collaborative ways to help defend the national enterprise, combining both the government interests and the business community. It defined the business side of this as a single infrastructure, with individual sectors using similar security systems. When I was on this board, Richard Clarke was the chairman. Howard Schmidt, a past White House cyber coordinator, was on it too.

At that time, sectors of the economy were to be treated as like-interest defenders of a component of the enterprise. The sectors were things like financial services, information technology, electric power, telecommunications, chemical industries, and surface transportation (such as the rail industry). We could argue that these groups have a common interest in protecting parts of the enterprise, and they probably have similar security issues to address, but they did not have much in common about how they did much of anything. Some of the utilities, for example, get power from other countries where they have very little say over how the infrastructure is protected. The same is true for most sectors. The CIP Council, the working body of this group, was heavily influenced by the financial industry, a large percentage of which was banking. It came closer than any of the sectors to understanding itself and its networks.

The financial sector saw its networks as in integral part of its business and had substantial regulation of financial transactions by the Federal Reserve. Very strong computer security policies were a tradition with the financial community. Most of the members agreed to them and favored information exchange about incidents. The financial and information technology sectors were pretty much in agreement that more had to be done to integrate the national infrastructure. In spite of heavy regulation, or maybe because of it, the financial sector has good policies that can be followed and understood by the participants, but this does not make its job any easier.

Richard Clarke had a difficult time getting any of the sectors to collaborate as well as finance and Information Technology (IT). They had much the same difficulty as the Federal Enterprise has with the different agencies acting independently. Part of the problem was the way they saw the threat. The financial sector is closely bound by threat—everyone is after the money—and they are all interconnected through the Federal Reserve. This is a tradition of very strong policies that are enforced, centrally managed, and inspected to be sure members meet a minimum set of requirements.

To some extent, the IT sector is similar because it provides services for commercial companies that outsource their processing and are connected to credit card services. The credit card industry has a similar strong policy but does not have the inspection authority that the financial sector has.

The rest of the community did not seem to act like they had a common threat, even though it was clear that information was being stolen from all of them. Getting an understanding of the threat across sectors meant we had to share information across them, not just among the members of each sector, as was being done.

The important thing that came from this was the idea that industry could share information about threats through a series of information sharing and analysis centers (ISACs), which have largely been unsupported since. Most of the ISACs worked, but they had several problems with the government side of the sharing process.

We had difficulty with more than one piece of shared information being inaccurate. In one case, the list of affected vendor models identified as vulnerable was wrong, and that particular vendor was, understandably, not very happy about that. I don't think anyone in any of the ISACs or on the committees thought this was a serious matter, but it didn't take long for the lawyers to express their concerns about their client's business reputation in some of their customer sites. The ISACs asked for liability relief so they couldn't be sued over a mistake like this.

For the next year or so the committee tried to get bills introduced, help draft legislation and persuade industry leaders to support legislation to limit liability for exchanges of information that identified vulnerabilities between members of the various ISACs. Sitting in those rooms where the mark-ups were being done was an experience.

There were all kinds of businesses and government interests represented, and it was hard to tell, by looking, whether a person represented a government interest or a business interest. There were bills introduced but none of them ever passed. Most of my associates saw it as the first step in sharing information about the software vulnerabilities of software vendors, something many lobbyists did not see as beneficial to their interests. It may seem strange that we could share interests in many respects, but not in identifying and sharing vulnerabilities of some of our members' products. The legislation took 15 years to pass.

The third problem with the government leadership was the Defense Department. Although it did not participate in the industry committee, DoD wanted to classify everything that dealt with any incidents that were being shared. What it typically did was accept unclassified reports from industry groups, add something to them that was classified Top Secret, then distribute any details only to the government, particularly DoD, and a few defense contractors with security clearances. This meant that people who did the initial reporting did not get anything to share from the government, and could not see anything they added to the report. It put the Defense Department and large defense contractors in the lead for cyber security.

The federal departments and agencies act like independent countries and not part of the same establishment we all know. I remember the general counsel of the Army telling us that we could not do security monitoring of army networks, even though we owned those networks. The army operated them for us. She said it was a privacy matter. That was a new definition of the privacy we used to know, but eventually, lawyers-to-lawyers, we were able to get this worked out. It was parochial thinking.

There isn't any reason to believe that we can't have one network in the federal government, as proposed in President Trump's May 2017 executive order. But that is not a new idea since the Comprehensive National Cybersecurity Initiative proposed, as its first sub-initiative, to "Manage the Federal Enterprise Network (FEN) as a single network enterprise with Trusted Internet Connections." So, the White House, during the Obama administration, saw the FEN as a single network, separate from the Internet. It may also see it as an entity to be managed, but it will not be easy to do because the Federal Enterprise is a mess. According to Personick and Patterson,

A General Accounting Office (GAO) report found that over 50 organizations (including five advisory committees; six organizations in the Executive Office of the President; 38 executive branch organizations associated with departments, agencies, or intelligence organizations; and three other organizations) are involved in CIP (Critical Infrastructure Protection). Adding in state and local entities would greatly enlarge the total number. As the establishment of the Department of Homeland Security in early 2003 underscores, the organizational structure of CIP—and within it, CIIP—may continue to evolve for quite some time, and the form it eventually takes will determine the extent to which infrastructure protection is singled out from or integrated within other elements of homeland, national, and economic security.[39]

The CIPB was eventually dissolved. It couldn't get much done even though it did seem to have the right membership. It needs to be replaced with something that has the power of the National Security Council to work together with industry and government. Its strength was in not being dominated by the military or intelligence community. It could work effectively with both, given national support. We need that type of leadership to come from the National Security Council, where the U.S cybersecurity coordinator manages the cybersecurity office. It is the only place high enough in the government infrastructure to manage the complicated political issues that arise between government and private businesses. Now that it has a chance of being a permanent office, it might even have a better chance of being successful. It needs to have representation from the business community and the federal enterprise at very high levels and set policy for the national enterprise. We cannot hope to defend the national enterprise without serious change in the management of federal networks.

Shadow War

I was program manager of a development program called SHADOW, an intrusion detection network that started as a thought of how missile defense might be able to do intrusion detection fast enough to find, stop, and maintain the information systems that make up a ballistic missile intercept network that finds and shoots a target before it can reach the United States. Missiles can get there pretty fast, so you don't have much time to fool around.

If we were going to stop a network attack it would have to be able to detect the attack event, identify the root cause, prepare to isolate it, and continue to operate the rest of the network to fire the defending missiles. Most of the systems we had were having difficulty doing this type of thing in less than a few hours. We had to be able to do it much faster than that.

What SHADOW showed us was something that scared a lot of people, including us. There are some pretty sophisticated people out there mapping our networks, testing various types of penetrations, and leaving behind little evidence that they had been there. The Chinese entry into U.S. systems without taking data follows this same line of thinking. Attackers were able to do some interesting things like this:

On Tuesday, a person pings a computer on a network by sending out a brief command, directed towards any computer that might be found at an address, that says, "Are you there?" Most computers will reply. On Wednesday, a person pings a computer on another network in the same subnet. On Thursday, another ping … and so on. If we did the same kind of thing on a street, each day we would mail a letter to one possible street address in a given series—like on the 400 block of James Avenue we send a letter to 401 and we keep track of the addresses where the mail is returned as "No such address." The second day, we send a similar letter to 400 Bluebell Lane, which is the next street over, and we keep doing this until we have all the street numbers. Electronically, this can go pretty fast, and at the end you would have a map of all the computer addresses of every computer on every network, if none of them were protected from such things. Nobody does this, of course, unless they are mapping the networks and don't want us to know they are doing it. They ping (or use a variety of other methods to get through firewalls to map inside) infrequently on any single network because anyone seeing this kind of activity on a single network would become suspicious. They were mapping all the systems where we had sensors, from the East Coast to the West.

Next, they would go back and run certain types of "probe" attacks against each system to check to see what types of operating systems were being used. Then they would try certain types of attacks to see if patches for known vulnerabilities had been installed on each one. At the end of all of this, they have a map of the network, what each type of computer is, what it is vulnerable to, and, if they take the time to update this now and again, they can attack pretty much anywhere and be successful. What they learned to do was to capture these vulnerable computers, chain them together, and use them to launch attacks against other computers. These people have a lot of time on their hands and they are very, very good at what they do.

It reminds me of something Dr. Parker used to tell us at USC: "Criminals spend as much time at their job as you do at yours." So do intelligence services. They were preparing to do successive generations of software builds on their attack software, each with new capabilities to do automated penetration and attacks and to gradually improve their products. We observed them testing but not deploying some software, which means they had capabilities they were not showing to anyone else. We were able to predict and warn certain people in the government that the attacks, which brought down eBay and a few others in February of 2000, were going to occur. We said they would happen in January, based on their previous software development cycle, but they did not keep their schedule up very well over the holidays.

This turned out to be a group of six people who had time on their hands and malice in their hearts. A government can devote far more resources to this type of thing, and they won't all take off for Christmas. The Chinese have already been accused of mapping the electrical grid of the United States,[40] but they think bigger than just mapping the electrical grid. What they probably have done is map telephone switches, computer networks, electrical systems, emergency management subsystems, transportation systems, banking and financial systems, and government. It is not that hard to do, but it takes time. Somebody has been doing it for 25 years now, and if we round up the usual suspects, China will be in there somewhere. It is something they would do if they are really interested in information war. They don't even have to use the capability; they do a couple of demonstrations just to let us know that they have them.

When they rerouted Internet traffic to China, we should have been paying attention:

> For about 18 minutes on April 8, 2010, China Telecom advertised erroneous network traffic routes that instructed U.S. and other foreign Internet traffic to travel through Chinese servers. Other servers around the world quickly adopted these paths, routing all traffic to about 15 percent of the Internet's destinations through servers located in China. This incident affected traffic to and from U.S. government (".gov") and military (".mil") sites, including those for the Senate, the army, the navy, the marine corps, the air force, the office of secretary of Defense, the National Aeronautics and Space Administration, the Department of Commerce, the National Oceanic and Atmospheric Administration, and many others. Certain commercial websites were also affected, such as those for Dell, Yahoo!, Microsoft, and IBM.[41]

Most of it was from our Defense Department, which says there is no reason to believe there was anything to be concerned about. It was probably an accident. This tickles my imagination because it just doesn't seem like something that happens as an accident might. Since it does happen in various parts of the world, on a regular basis, it is possible. It is also possible it was just a practice for something bigger.

The Second Principle of War

The most chilling thing Von Clausewitz said about war is something you have to read more than once to absorb: "for in such things as war, the errors that proceed from a spirit of benevolence are the worst…. This is the way in which the matter must be viewed, and it is even against one's own interest, to turn away from consideration of the real nature of the affair because the horror of its elements excites repugnance."[42]

In other words, war may make you sick to your stomach, but if you are going to fight one, it is better to do it without thinking about how ugly it is, or might become. This is something nice guys do—they turn away from it because it is ugly and vicious and they don't like to do things that are either one. We won't win any wars that way, and we certainly won't win this one.

The Chinese are using a simple strategy to get access to the rest of the world's information and control what they can't get. Their Second Principle of War has morphed into "Own it; don't attack it." It is better to buy into an infrastructure than try to hack into one. They can live inside infrastructures they own and don't have to worry about whether someone finds them. They are supposed to be there.

When the deal between Huawei and 3Com fell through, it didn't take the Chinese long to start working on another purchase that would put them into the U.S. markets for telecommunications. Huawei was already selling equipment for its networks to Cricket, Cox Communications and Clearwire, later bought by Sprint. They are working on chips and telecommunications companies in the next round.

Opportunities, Opportunities

As an Apple user, I have to admit Foxconn makes a good product that I like, but it makes me nervous. In a war, Foxconn has access to almost everything I am on the Internet because they make it—iPhones, motherboards for other computers, iPads and lots of other things too. People who make things have the most access to the internal workings of the product, and they can modify that product in ways that would expose the users to hacking that cannot be stopped because it is built-in.

Someone, a little more careful, can build in hacking software or firmware, and do it so they wouldn't get caught. People after ATM and slot machines have done all these things at one time or another to make money. People who gamble seem to have great ideas when it comes to getting into slots, including the first of the known hardware hacks where a complete circuit board was replaced with one the user controlled. Slot machines are just another kind of computer. That was 40 years ago, for you budding hackers, so it is not so easy to do these days. Getting caught in a case like that means going to jail, but this is a little different. When a company intentionally making modifications gets caught, it can be really bad for business.

It has a huge cost in product acceptance. How many iPads would you buy if you found out they were transmitting everything you printed to somewhere in Estonia? If we were to find out that Foxconn was building a back door into every iPhone it made, then would that have an impact on iPhone sales, and on our ability to trust Apple products in general? I trust Apple, but Apple does not build the iPhone. I want to know that those who do build anything I have on my computer network are good guys—or at least neutral.

Lenovo, the world's largest maker of laptop computers, is owned by China, which bought it from IBM. Their computers phone home periodically to update software, and we don't think very much about this. All computers phone home and all can download most anything. It is a similar type of opportunity. Every major computer manufacturer has some of its computer equipment made in China, and Dell has three large manufacturing facilities there. Even some companies building computers in Taiwan have manufacturing in China. If we look at the range of computers and computer equipment made in China, that risk could be bigger than we might like.

Governments have taken to calling this the "supply chain problem," but just making computers is not the real problem for the rest of the world. The Chinese make the root components of the networks all of us use, the main parts of the networks interconnected through every country. 3Com was making them too, when Huawei was trying to buy them. If they start putting backdoors and hacker access into those boxes, or they manage the networks they ride on, they can restrict the Internet access we get. It isn't easy to do, unless there is enough scale in the attack. The Chinese sure seem to be building to that scale.

The Internet is not one thing. We don't really notice how it works when we access it because it seems like everything is just right outside our door. That makes it seem simple. Go to the Internet Mapping Project, http://www.cheswick.com/ches/map/, and look at how Bill Cheswick, who has been mapping the Internet since mapping was popular, shows the layers of Internet service providers that there are in the world. Verizon is in there somewhere, but it looks so small. There is a maze of service providers that is so big and interconnected that almost any kind of attack won't be successful across all of them.

But I want to take you back to a point I made early in the book about all the components the Chinese make. The Chinese seem to think big, but start slow. It is possible they have already managed to get counterfeit chips into fighter planes in the U.S., and 400 fake routers into our networks.[43] They certainly have fake circuit boards and wide area network cards.[44] These have been discovered because some of them failed and the owners complained to the people who they thought were making them. With time, they will get better and won't be detected quite so often. They have access to almost every kind of computer component they make. They have crept into this market with fakes of various types, components and whole routers. It will only take them a few years to get those parts working the way they are supposed to. By that time, we will have a whole lot of these things in our networks, in places we are going to be sorry about.

There are new kinds of viruses that can redirect my router connection to some that are controlled by whoever is passing this thing around. The Chinese certainly know this exists, or they might have invented it, since they have so many antivirus efforts going on. What this allows them to do is route traffic to networks they control. Then they may not have to control all the equipment on the Internet, to have everyone connect to something they own.

With every passing day they have new opportunities, and they are expanding those as fast as they can. In those five network companies the Chinese own, there are circuits all over the world, so they control huge portions of the networks that use them. They supply most of the network components and phones from supplies they make. They have agreements with other companies that give them access to more. What's more, these companies are government-owned. They don't say very much about themselves on the Internet. Most of them are based in Hong Kong and they have no competition. No international companies may compete with them in China, so nobody is going to buy into their networks. When it comes to global infrastructure, the Chinese own a lot of it, and we may not realize how much, or what it means to us.

China Telecom is the oldest and largest of their telecom structure. It is the largest mobile phone company in the world, by number of subscribers, and has the largest fixed-line network. Its leaders are members of the Communist Party first and businessmen second, similar to the Soviet system of the '80s. It owns circuits in China, Japan, Central and South America, the Middle East, Australia, South Asia, Europe, and the United States. The government spun off China Mobile and China Satcom to help the growth into these markets. China Telecom still has all the fixed land-lines under its control. Their undersea cables are in Hawaii and several places on the West Coast.[45] So while they may seem like companies that just operate in China, they have arms with a long reach.

China Mobile is the largest mobile telephone operator in the world, having 70 percent of the domestic market. With so many people, they quickly get to a high number. They are 74 percent owned by the government, though as we saw after replacement of their CEO, 100 percent controlled by the Party.

China Unicom is the only state-owned telecom to be traded on the New York Stock Exchange, except that its two largest owners are both state-owned, so it is a little difficult to think of them as public companies. It is the second largest telecom company after China Mobile. China Unicom and the Spanish telecom Telefonica are combining investments that they claim will be 10 percent of the world's market. Unicom gets a seat on Telefonica's board of directors. Through a separate deal, Telefonica and Vodaphone are sharing infrastructure in Europe, putting Huawei, which has separate deals with both of them, in a better position for expansion.

PCCW Limited, together with its subsidiaries, does telecommunications services primarily in Hong Kong, on mainland China, and in the Middle East and the Asia Pacific regions. It offers local, mobile, and international telecommunications services, Internet access services, interactive multimedia and pay-TV services, plus computer, engineering, and other technical services. The company gets into investment and development of systems from offices in Hong Kong, mainland China, the Asia Pacific, and the Middle East. While U.S. and foreign companies operate in China, they are not buying into its networks, but China is buying into everyone else's.

Because of the way international business has merged over the years, most of companies that own our communications systems are no longer just U.S. enterprises. We still have rules that limit the amount of ownership a foreign group can have, but limits do not mean none. Just as simple examples, Vodafone from the U.K. and Verizon are teamed in Verizon Wireless; Vodafone owns part of a French telecom company that is being bought by Vivendi, a Paris-based company; T-Mobile is a German wireless services provider, owned by Deutsche Telekom; the Alcatel-Lucent Technologies merger produced a company that does business in 130 countries and has employees with 100 nationalities. Acatel-Lucent, with its headquarters in Paris, is still ahead of Huawei in selling communications equipment. They include Bell Laboratories, which did much of the original research that forms the underlying telecommunications infrastructure. Some of the large telecoms are government-owned or have substantial government control, so they are not much different from China.

The router that I use on Verizon is made in China and supplied by Vodaphone to my Verizon FIOS connection. As unhappy as I am about that, I am having trouble finding a router made in the U.S. The Chinese have the market locked up. A *Washington Post* article says NSA talked AT&T out of buying some equipment from them too; we already know about the Sprint/Nextel deal. I want to buy the router NSA buys. There is just too much going on in the world of business to keep up with it all, but it certainly is something that bears watching a little more closely. Routers are too important in directing Internet traffic to be left solely to the Chinese.

If they can't get to me directly, they can find another way. Vodaphone opened a joint research center in Italy with Huawei. Huawei just got a contract to replace 8,000 wireless transmitters in Australia, on a Vodaphone contract. In a few years, all the 2G and 3G phones will be running on Huawei equipment, with a Huawei 3G phone to go with it. They got a 5-year deal to do managed services for Vodaphone's Ghana operations, and this relationship is just getting going. It is difficult to see exactly who you are buying from when the marketplace gets so complicated. The French company Vivendi is buying back its shares in French mobile carrier SFR from Vodaphone. That seems to look like a smart move. The French are careful about things that affect their national security.

Other governments and big businesses rent this infrastructure from the 58 major telecommunications companies, the same way an individual does: they buy service from them. What they don't do very often is control how these services are protected. So, among other things, I am happy with the way my Blackberry encrypts e-mail because I cannot rely on any of the vendors who sell me service to protect my e-mail in transit. They would say that was the customer's problem to deal with. From an infrastructure standpoint, the vendors would argue that they only lease circuits, and the consumer of the service has to protect it from other people who might use the same service. This logic will not help those who do not have service if someone takes it away. I can use an example that makes practical sense to anyone, regardless of where you might work.

When I worked for the government, we had a contractor that was supposed to be designing a network that would be used to connect parts of a military command network, and it was going to put an infrastructure into place to start this work. The contractor wanted to lease the circuits from an undersea cable company that was owned by the Chinese and based in Hong Kong. From a security standpoint, it did not seem like a very good idea to have a cable that had actual connections to Mainland China and Hong Kong be used for such a critical function. Some might argue that if we can do banking this way, we surely can do anything else. We told the contractor this was not to be done and gave them the main reasons: the company was foreign owned; we did not trust the Chinese all that much; and we didn't care that they were cheaper. This news did not even slow them down.

The next step was to bring in a defense security service specialist in foreign ownership, control and influence. She explained the ownership of the company in question and how it was not wise to have a component of a command and control system riding on a network owned by another country, the Chinese aside. Having a Chinese owner made it impossible to consider such a move. They were not paying very much attention, or so it seemed, because they just kept plodding along towards that connection.

I went to see our general, who was a pretty sharp guy and saw what was happening. He said to bring it up at the staff meeting of our senior officers and he would take care of it. When it did, the general turned to the project manager with a look that would melt anyone who could look back, and said, "Is this true?" The contractor started to say that there were some very good reasons why we should consider buying these circuits, but he had not gotten halfway through the argument when the director held up his hand. "Stop," he said. "I can't believe you were even considering doing this and I don't want to hear any more about it." There was no argument from anyone, and the meeting moved on to the next topic. They leased the service from a U.S. company.

I might have felt better if this was the only time a U.S. defense contractor ever put its business interests ahead of national security issues. We are going to regret not controlling this type of activity. If the Chinese escalate their information war, things like this will matter. They will have control over our networks and can deny the use of them.

These services are like my FIOS connection to the Internet. If someone takes that from me, my data is still safe on my own computer, but I can't use the network to send e-mail or search for other data. In the case of war, I could have the military orders interrupted by having someone deny service to the network. A group of hackers that is mapping every computer in the United States is thinking about how to attack them all at once to deny sectors of the economy or military the use of those networks. We will still have our data but the infrastructure we need to use it will not be available. All the Chinese need to do is shut off those connections on every machine. And they don't have to do it; they just have to demonstrate that they can. They can win without firing a shot. Sun Tsu thought that was a good idea.

So, with all the communications equipment and circuits the Chinese have, they are close to being able to disrupt quite a bit of traffic in the world. They won't want anyone to see that, of course, but they are ready if the need arises. Before that happens, they will start trying to get people off of their protected networks and onto something they own or control. There are really two ways to do this kind of thing, besides the virus redirect I mentioned earlier.

The first is the "let me make you an offer you can't refuse" way. Make it cheap to buy into. They subsidize their vendors and teaming partners to make it cheaper to have their service than that of their competitors. They give low-interest loans and sweetheart deals to attract customers. They already proved they can do it with solar energy. It is really hard to say no to a good deal. They sell these services at the shop on the corner, using vendors we already know and use. They don't make a lot of money, but they do all right. If the companies are big enough, they can threaten smaller competitors to get with their companies. I haven't seen any instances where China has done this, but they certainly have the market power to do it.

The other way is to make the other services less attractive. If the Android operating system is free and can be used anywhere, try to make sure it works better on your phones than on anyone else's. Apps cannot work as well on other types of phones as on your own. Conversions of data cannot be as accurate and complete. Web interfaces cannot be as smooth. The idea is to intentionally influence that without getting caught. To do that, they have to be able to replace the operating system supplied by the original vendor. If they were the original vendor, it will be easy. If not, it can still be done, but it requires much more work.

If RIM has business servers that are really secure, an adversary might want to go after services connected to it and see if there was something to be done to make their products not work the way they should. They could make them look less attractive. That usually makes the other things on the market look better, even if they aren't.

There are some software vendors (you might remember the browser battles a few years ago) that have been accused of this now and again. Users say their data didn't convert quite as well with one browser as with another. Imagine that. They called it "enhancing the user experience" in those days. One after another vendor was trying to say their browser was the best, and the measure was how well it worked with other applications those vendors used. Some of them finally went bankrupt trying to keep up, and the market settled down a little. Driving out the competition is good for business.

Believing in War

There have not been very many wars where one country has opened up talks by saying, "We are going to pound you into dust and take all your territory. From then on, you will do things our way." Those are fighting words. It is never that simple. Usually the future combatants will start off by saying how great and wonderful everything is between them and how much they need each other. When enemies say that, there is trouble brewing. We aren't at that point with China just yet, but we will be.

Both Secretary of State Hillary Clinton and her deputy, Williams Burns, spoke about U.S.–China relations as "challenging" and "sensitive." We could say that about Israel, France or Germany some days, so that doesn't tell us very much. They said we are concerned about China's military build-up and their "incessant cyber attacks on public and private American entities." They were not happy about "bilateral economic priorities," which is State Department speak for trade imbalance, nor about the Chinese ability to try to protect intellectual property of our businesses. And, of course, there is that little matter of the currency controls that are causing us no end of grief. They think we should talk more, and they have set up some chances to do that with strategic security discussions. Considering the source of that, it is not much of a surprise that talking is always the best thing anyone can do for future relations. More talk (frank discussions) means more trust, the way the State Department looks at things. State always wants to talk, and the Defense Department wants to send ships. Neither of those will work very well.

The Chinese would be the first to say that they are just commercial people, trading with the rest of the world. They are doing things that every other country does and doing it better than the rest. They are not at war with anyone. There are plenty of Henry Kissingers of the world who really want to believe them, but those State Department folks are saying we should talk and they are saying it with the background of some clashes that are starting to concern them.

Most rational people do not want to be at war with anyone, but they also know the difference between war and *not war*. Sometimes it is just a matter of intent. Aleksandr Solzhenitsyn, the Russian writer, said: "If only there were evil people somewhere, insidiously committing evil deeds, and it were necessary only to separate them from the rest of us and destroy them. But the line dividing good and evil cuts through the heart of every human being."[46] It isn't about whether they are nice people or not. It has to be measured by what they do. They look like the devil to me.

The Chinese really believe they are the world's strongest nation, without having the most powerful military. They haven't been growing their military steadily, but it was not a priority for them. They have been growing their businesses, especially those that related to networks of all kinds, because they believe control of information will help to equalize things between countries stronger than they are. How many countries are stronger than they are? Just one.

They have a centralized management that can direct how they build themselves, but they have a long way to go to become the type of world power that we are. They buy up the world's communications systems and put deep roots into them. They hack everyone and they steal business secrets from everybody, not just us. They have the ability to control their Internet but that hacking continues like it is part of the accepted practice of the government that owns the capability. They are trying to get their state-owned companies into every network the world has. They back off when discovered, and try again a different way. They spy, like everyone else, just more often. Over time, they have gravitated to this kind of war because it is more successful and less dangerous than the alternatives. They like it this way, because they are winning, and they are glad to talk, stall and delay in any way they can.

If you go back and read the definition of information war, the way RAND laid it out for us, they are doing everything by the book. They may have changed it a little to fit their culture and way of communicating to their army, but it is pretty close. It looks and feels like war, though few call it that.

Deterrence

It would be nice if they couldn't get away with this sort of thing, but we can't just say "Stop" and expect them to pay attention. We have to make it more difficult to continue. Since they are not going to help us out by telling us what to do, it is not as easy as we would hope.

Talking might help with the economic warfare, because the European Union and the U.S. are not the only entities in the world that are behind in their loan payments to China. There is enough resentment to get something going to put pressure on them. That is the kind of thing the State and Commerce departments can do. But they can't do much about Chinese intentions in the cyberwar and space, as in outer space. These are things that need deterrence to slow them down.

Deterrence is a kind of threat that something bad will happen if the behavior isn't changed. If my dog nips my hand, I smack him with a newspaper. If it happens a few times, I don't have to smack him; I just pick up the newspaper and start looking at dog ads. He behaves without having to be hit. So far, the Chinese can thumb their noses at anything we can say we will do to them if they don't stop. We don't have a good deterrent strategy for China or Russia, which would spill over to North Korea and Iran.

They are like a big bully who is at the bus stop where our neighbor sends her children. She can drive the children to the bus stop and wait with them until the bus comes, or she can call this boy's parents and talk to them about his behavior. She can call the police if some violence is done. She can call the school because in this part of Virginia, children can be disciplined in school for what they do waiting for the bus or getting off the bus and going home. She can train her child to fight. She can hire a guard. These are all things she could do, but sitting in the car at the bus stop works, so she hasn't looked for another alternative. She has a deterrent. The bully knows she can get out of the car and stop his behavior because she is bigger than he is and has some status. He would look bad hitting a woman, so he can't really do very much.

Deterring a country is harder than stopping a bully, but some of the same principles apply. The most important is that the threat has to be credible. I wanted to go over to the bus stop and threaten the little brat with bodily harm, but that is not a credible threat. It might be a crime, and I'm sure the little guy can read, and knows it. Those mothers sitting in their cars would not let that happen, either. One of them might be his mother, but I doubt it. It wouldn't matter. They will defend any of those kids, even the bad ones.

Our leaders seem to think that talking about this will turn the Chinese around, but that is not going to help. We have to pay attention to them. The White House might remember Norm Augustine, the CIO of Martin Marietta, when it merged with Lockheed. He said it was not so important for an executive to do something; they just have to pay attention to it and the right kinds of things will usually happen. We need to start paying attention to what they are doing and what can be done to stop them.

Having someone else have a chokehold on the world's telecommunications is not what we thought about 20 years ago when the military was planning for information war. We owned the Internet then, and many people outside the U.S. still think we do now. Not true. The shoe is on the other foot and it hurts, but it doesn't hurt enough. We have to have more interest to stop the kind of things the Chinese are trying to do, and we have to believe it is war.

We thought we could get their attention with some trade sanctions and some letters to the WTO, but that didn't work very well. They need a dose of their own medicine; and that would be having our government share limited amounts of intelligence with businesses trading with China. We have a really good intelligence community, and we don't use it very much for the kind of things that will help us here. We have executive orders that prevent it, but we need to think about changing some of those. The intelligence communities need to be involved in discovering where those abuses are taking place and countering them through special cyber operations.

It is not a secret that the dictators of the world are squirreling money away. Look at what happened when Gaddafi and Mubarak were missing in action. Their money was being "assessed" by every major bank in the world. After the Middle East settled down, everyone started following the money trail. The new countries wanted it, other governments wanted it; banks wanted it too. We have made it illegal to give these leaders money, and the rest of the world probably thinks that is funny.

I would like to know who is producing counterfeit goods and where they are being sold. I can't stop China from selling them internally, but I can stop them from being sold outside the country, if we start focusing on it and giving resources to people who do that sort of thing. Let's spend some money trying to discover or stop it. The Chinese can have all of those counterfeits they want. They fall apart in a few months, so they deserve them.

Our intelligence community has a great amount of talent for reverse engineering things. It would be nice if they could apply some of that to identifying stolen trade secrets being incorporated in Chinese goods. It would be nice if they could find some fake systems or some of that software going into our infrastructure. We would have some real information to give the WTO then, and it can be used to sue U.S. subsidiaries of some of those companies.

Our national business leaders naively believe that we can "out-innovate" China by just doing what we have always done, but checking the number of new research facilities in China and China's teaming with various researchers outside the country, that is not very realistic. We are selling them the ability to compete with us now, and in the future.

I want to know more about those PLA businesses operating in the United States. It needs to be harder for them to operate here—much harder. We can do what they did to Google and shut off their power every now and again. We have lots of trouble with power anyway. I don't like the idea of them being allowed to operate here, and want to make sure I don't buy a washer from one of them.

But the worst problem we have, and the one that is most difficult, is hacking. We could try stopping their hacking by jamming their sites, using logic bombs and Trojan horses or any number of other things to disrupt their hacking networks. It seems like this should work, but it never does, because the hackers use legitimate sites to store their attack software and data they have retrieved. They are not hacking us directly. We might be attacking some furniture company in Iowa or a clothing store in my own town. When we find them, they just move to another place. Hackers understand deterrence as well as anyone, and they like to avoid it.

Every president in the last 25 years has said we need better-trained people to handle computer security, when what we need is less security and more deterrence. Nobody in the government has figured out how to do that yet.

We can broaden our diplomatic effort to see if we can appeal to the federal government to talk to the Chinese directly. Before you laugh at this, it works once in a while with criminal gangs stealing money or information from more than one company, though China is not a country that cooperates very much. I was occasionally surprised by how much cooperation there can be between countries on criminal matters. Our law enforcement has even gotten some help from the Russian government and most of our allies, but it does not seem to work with China, Iran or Slovakia. There could be a few reasons for this, but not any that would favor China, Iran or Slovakia's image.

We can spend a lot more money on security of our systems and try to keep people out by making our target harder to get to. This is the equivalent of having a guard sitting with the children at the bus stop, only it doesn't work nearly as well. It used to be the way of business people everywhere, using the philosophy that you don't have to run faster than the bear, you just have to run faster than the person with you on the trail. That doesn't work anymore. Now that they can attack everyone at once, nobody is very safe. It makes the board of a corporation feel better, but it has very little deterrent value. The hackers know they can get in.

We are running out of options here, for a reason. There is almost nothing that deters this type of activity, especially where the government cooperates in protecting the people doing it. For trade, spying and hacking, the rules for deterrence are basically the same. If we are nice, there is no deterrent.

We tried to play nice in trade and it got our trade deficit raised every day by Chinese currency manipulation. In hacker circles, we publish the list of Internet addresses where these people operate from and they move their operations. We block them and they try a new method of attack. We can neuter them sometimes by modifying their software but that only works for a while and a new version is out that works. Most businesses are too slow for that to work.

The real problem is there are ways to deter trading schemes we have seen the Chinese use, and we can deal with hackers by making their lives more difficult and painful, but it requires our government to target them and undermine them. They are not willing to do that. Too many lawyers tell them they shouldn't. We need some new lawyers.

In the case of this type of hacking, I have to agree with that Chinese general who said the only way we can deter such a thing is to have those capabilities ourselves ... and more of them. We need to increase our attack forces and turn them loose. Find out what our enemy is doing and how they are doing it. Bury ourselves deep inside those operations and head off their plans before they can get us surrounded. The world saw how hard it was to deal with a group like Anonymous, which is not very big and is not backed by a government. This is a much bigger operation and has been going on longer. We can stop building fighter planes and tanks for a few months and start building up our computer forces.

We should help industry too, but not the way we do it now. The Defense Department seems to be able to help defense contractors by giving them classified information about the attacks against their computer systems. They need to start giving it to anyone who is being attacked. Defense used to classify the sources of attacks that were occurring in industry and then only give the summaries out to companies with clearances. Oftentimes, they were denying that information to the companies that reported it to begin with. It made no sense then, nor does it now. We spend too much money on collecting the information about who is hacking us, then never give it to the people who need it.

We don't have the stomach for information war. We have to go after the people on the other side of this with a vengeance. Attack them. Disrupt them. Infect them. As long as we don't, they win. If we don't stop them now, we have a bigger war to engage in. it will be much easier to fight them now, then wait for their successes to make it more difficult in the future. They are not unwilling to use real war if they think they will win.

Almost Real War

Von Clausewitz reminded everyone that wars should be fought without regard to how bad it might be. The Chinese and Russians seem to be considering war in a way that crosses a line between information war and nuclear war. That is not a fine line to most of us.

The person who invented information warfare in China wrote a book called *World War: The Third World War—Total Information War*. This is a long title for a book, but the Chinese characters make it look shorter. His thoughts he expressed are shorter too. He was concerned that China is vulnerable to information war in a slightly different form. He talks about those with the weapons of war, whether computers or nuclear weapons and how they have first strike and second strike capability. It always amazes me how military people can talk about mass destruction of millions of people like it was an Xbox first-person shooter.

The Russians and Americans always talked about this in the context of nuclear weapons and who would use them first. Both of them said they never would, but they both had them. Since the Chinese do too, we already know that game pretty well. They say they will not use them first, but they are undecided about when a first strike might be necessary. We say that we will not use them first, but we keep submarines out in the ocean with them, just in case. The Chinese have nuclear submarines with missiles on them too, so that part balances out pretty well. Numbers are not so important when the weapons are nuclear. A few can go a long way towards deterring one another. If the Chinese want to deter someone from launching a first strike, they have to have the weapons to launch a second strike. This is called first strike deterrence because it keeps the other side from thinking about launching one.

This is not as simple as it sounds, because the Chinese view of what might be a first strike might not be the kind of war we are thinking about. If we go back to that Chinese general who said he thought we might just throw up a nuclear weapon if the U.S. decides to break out weapons in defense of Taiwan, he was not talking about dropping one on Los Angeles the way we dropped one on Nagasaki; he was talking about shooting it off in the air, high up. There is no nuclear blast incinerating houses like we saw in those training films made in the 1950s, or a large fireball sprouting up out of the ground. There is just a flash and nothing. Well, "nothing" may not be the right term, since it is really nothing anyone can see. It is something called an electromagnetic pulse (EMP), and it is not very nice.

An EMP can do in one minute what nuclear bombs can do, but they don't leave such a mess. They don't kill very many people, but they are hard on power lines, automobiles and telecommunications circuits. Nothing that has a circuit will work unless those circuits are hardened against EMP. We can think of being without electricity for a while, but it would not be fun to be without electricity, cell phones, cars, trucks, portable computers, circuit breakers, back-up generators, water purification, ATMs, banks, and most battery operated emergency devices. In a published report a few years ago, a congressional panel thought these effects might be seen in a circle of around 700 miles.[47] That would be a long way to walk to the grocery store, which will be empty by the time we get there.

The only good thing about this type of attack is the deterrence value of having the same type of weapon in your own inventory. That would certainly make China, Iran and Russia think twice before letting one of these go, but North Korea will be on that list next year and they don't seem quite as stable as the rest of us are, or dependent on the same types of technology. A country like North Korea might get along fine without electricity or any of those other things. They may not have them now.

Nuclear weapons are usually a sure sign of war. When someone starts setting one off, whether it is high up or not, it is going to cause some real problems because the next step is for us to do the same thing back to them. If North Korea sets it off, then we are tempted to do the same to them. The Chinese would say it is a shame that they can't control North Korea, but they are our friends in a strange sort of way, and we will protect them if they are attacked. It is a tricky situation, but the people in Los Angeles want something to happen and they are tired of eating food out of a can. We hit those North Koreans with something they can remember for a year or so; then we have to deal with China.

Once that starts, it can be difficult to control. That first explosion detonates over Pyongyang, and then what? Do we both just sit and wait to see how that worked out for us? Remember that the comments of the Chinese general were related to Taiwan, so in the meantime, the Chinese are over-running Taiwan, which is not very big; it would not be long until that was over. We might not even see China as the opening round of this war. None of the options are very good, particularly for us, and even China probably does not like the scenarios they have put together for this type of event. Radiation is not very pleasant for any of us.

The Chinese want Taiwan back, and they think it was given to them after World War II. They think the U.S. and some of its allies are responsible for it not being given back. They just would not want to start throwing nuclear weapons to get it. But the Chinese general mentioned another type of thing besides nuclear weapons: viruses.

Worms and viruses don't make a mess and they are not usually seen as war. They don't produce radiation or burn up children. They are clean, so to speak. We have even managed to get a virus in our Predator drone systems, and I can't imagine the kind of idiocy that allowed that to happen. You would think, with the importance of this weapon to killing off terrorists and doing surveillance of the ground in Afghanistan, that our military would be a little more careful with it. What are they thinking? They are lucky to still have a drone system to work with.

These are not the kind of viruses done by kids who are using a virus kit that they got off the Internet. Those are known viruses, and the antivirus companies spend quite a bit of time keeping track of the development of them. The general is talking about combat viruses, and these will not be float-ing around on the Internet waiting for someone to figure out how they work. This helps them be a little more controllable, but nobody knows how con-trollable they will be. We do know they are a good sight better than a nuke.

Symantec did an analysis, by country, on where the Stuxnet worm showed up. (Just as an aside, after CFIUS overruled a takeover of Symantec, the antivirus company, it formed Huawei Symantec Technologies Co., Ltd. Huawei is the majority partner with 51 percent ownership, with the business headquarters in Chengdu, China. That agreement has ended, and the partnership dissolved.)

Though not in large numbers, the worm got into systems in India and Indonesia, more often than it did in Iran. Since these are reported incidents, the Iran number could be a little short of reality, but we will never know. Pakistan, Indonesia, Afghanistan, the United States, and Malaysia were all places where it weaseled itself in. Now, it appears that this particular worm only attacked the control software for certain types of equipment, but the software was used in more applications than just centrifuges. It is more difficult to control things like this than to talk about them, because the unin-tended consequences are not nearly as easy to see in a laboratory where they are built. Also, the effects are not nearly as predictable as the developers think they will be. Software developers think their software is always perfect, even if the Internet is not. Things will happen that they did not anticipate. They will say, "Oops, sorry," but the driver of the car that crashes into the train is saying a good bit more.

The worm that propagated in March of 2017 was linked to ransomware; it encrypted computers so they could not be used. That attack infected just over 300,000 users in the first few weeks and is probably lying dormant in many more. It will not activate because the machines were patched and it can't get in. So, suppose we have a worm that is much more extensive and focuses just on the electric grid, or a good part of it. With the power out, not everyone can do a patch and will be doing something else besides using their cell phones for everything. Doomsday, you might think. This will cause quite a bit of intentional damage that will cause us to start thinking our national security is at stake. We tend to roll our nuclear subs and airplanes at times like that, so it isn't something someone is going to do without a lot of thought.

Our government will want to hunt for the country that does this. There will be some unintentional damage (sometimes called collateral damage) to clean up, because more things are on the electronic grid than war planners tend to think about. We saw this when hurricanes brought down power lines to large patches of land. That fellow across the street cannot live much longer than the 4-hour battery life on his dialysis machine. My mom says the food will last a few days in the refrigerator after a Florida hurricane. Hospitals have quite a few people who require constant care, and many of the hospitals will eventually run out of power. Knocking out the electric grid will put everyone in the dark, with no street lights or traffic signals, making emergency calls more interesting. If we don't have TV, there will certainly be a revolution.

The police forces are pretty busy, but crime goes to nothing. There is a good trade-off there. We won't be eating out as often, but crime will be down.

Emergency generators will work for those who have them. The rest of us will not be eating quite so well, *and* not eating out. I will get testy after a week of eating out of a can, so I may lead the revolution. Guns may come out, and we may need some police in my neighborhood.

Getting fuel for generators and cars is a little more complicated, but it may be possible. There is a manual pump that is available for emergencies that would allow people to pump fuel out like old-fashioned well water. Most cities have backup power for sewer and water, but rural areas may rely on batteries that won't last forever.

The real problem is that the electrical grids are not all in the U.S. and they are interconnected, so someone who tries a virus will find it in places it should not really be, even in some allied government's grid. The Canadians, Brazilians, Central Americans and Mexicans will not like having that virus in their systems. It could get into international grids and turn out to be harder to control than those folks who worked in the lab thought it was. Oops. This would be annoying to the leadership. Global war is complicated, and just as complicated and risky for the Chinese as us.

Eventually, we are going to figure out that it was a virus and start working on a solution that will reduce its damage or get rid of it. This might take a few weeks, or less, if we are lucky. Some portions of the grid might not be affected. Maybe we can figure out why that is and fix it for all the places. These types of things do not last forever, so using them is not something a person would do without quite a bit of thought. Figuring out who did it will be possible, but it may take months. It can ultimately be traced back to its country of origin, maybe to a place. Then, we have to figure out what to do about it. That is the hard part.

The nice thing about lining up armies to fight is that we usually don't have to wait to figure out who it was that we fought. With viruses, that takes time. At Pearl Harbor, it was pretty easy to see the meatballs on the side of the airplanes coming in to drop bombs on our ships and people. This would not be like that. I think people are confused when they say there could be a digital Pearl Harbor. It won't be that simple.

After we answer the question of who did it, the next obvious question anyone in this situation has to answer would be, "Is it war?" This will be important to all those soldiers and sailors on nuclear submarines that are moving toward Southeast Asia by that time. Some people have published reports saying that the Chinese and Russians have already gotten into our electric grid and planted software that makes it easier to come back and do more, potentially to disrupt operations on the grid.

If both the Russians and Chinese got into our electric grid and planted software to get back into it and disrupt it, and actually did it, to prove that they could—would that be an act of war? Probably not. We would like to think it was, but it is like the gunman who holds up a person on the street with a finger in his jacket pocket. You couldn't charge him with armed robbery with a finger. If we believe he has a gun, we might defend ourselves, right then, as if he had a gun. The robber has to run that risk. The Russians and Chinese would rely on us knowing that they planted the software but didn't do anything that would cause us harm. We have to know that they are doing this for some reason that looks like war, and that is as close as it gets to war, without pushing things over the line. That is just how the Chinese work, pushing us right up to the limit to see what happens.

For those who say the Russians and Chinese did this, we have the little problem of attribution again. The software might have some programming that we can recognize from somewhere in China, but that won't prove much. Can I say that because the attack took place from a city in China that the government of China actually was doing it? If a criminal gang did it for extortion, do we blame China for it? If North Korea did it, would we blame China? Then what do we tell our guys in those nuclear subs? "Wait, while we figure this out."

You can go to war with a group of people in another country, as we have done with al–Qaeda, but they usually have to do something to you that justifies that. China would not like us fighting with people inside its borders, even if they were doing some terrible things. It is a much harder problem than just declaring that someone is doing something bad and needs to get whacked for it. So do we have to wait around for them to do something really bad to act? Only if we want to go to war. As you remember, nobody goes to war these days, so let's not.

9

Drifting into Darkness

We have always believed that ordnance on target wins wars, so we have some really big weapons, such as aircraft carriers, as big as a city of 5,000. We build lots of expensive airplanes that can carry bombs or shoot down other airplanes, and we think those will help us some day. We have some experienced fighters after Iraq and Afghanistan, and whole shiploads of Strikers and personnel carriers. If we haven't given them away, they might be useful for the wars we learned to fight.

We know where to put a bomb or how to get it there in one piece and make sure it doesn't kill a whole house full of innocent people, and the idea of the bomb is important to winning wars. One of those al-Qaeda in Iraq, or Taliban leaders, can understand a bomb on the front of a missile, strapped to the wing of a Predator or Reaper, as easily as we understood the plane striking the World Trade Center. We see ourselves at war with al-Qaeda, but it is harder to see that we are at war with China.

If you saw the fight that broke out when Georgetown's basketball team went to China to play a Chinese army team, you saw just a brief glimpse of the feelings involved. It is easy to dismiss it as the heat of the moment in a sporting event, but the look on the faces of those army team members as they were kicking and hitting the Georgetown player on the floor gave me the feeling that there was more to it than just a basketball game. There was real hate there. They were frustrated and they were not going to take it anymore. Somebody in that army, maybe above the army, was steering them in that direction.

We have to believe that too much power has been placed in the governments of dictators. The military influences how the economy expands and how the civilian populations are managed. There were indications in the case of the Queensway Group that the Chinese senior leaders were trying to put distance between themselves and some of the company's activities. They found it hard to do. Even the planners know they are a little out of control. That is dangerous and has been a point of reform by Xi ever since. He faces resistance because he is tackling the army and the army's corruption at the same time.

China's military is not friendly to the U.S., and "not friendly" is not really descriptive enough of the feelings. The secretary of defense said the display of the J-20 fighter was not something President Hu Jintau seemed to have been aware of, meaning the military thought it was useful to use the secretary's visit as a show of force, and may have acted alone in doing it. There are also divisions between the army and the police, who run the border patrol functions; this means the sea lanes and fishing rights are being enforced by non-military forces.[1] They are in a constant struggle for influence among themselves and with the Central Committee. They can criticize military exercises of the U.S. in front of the chairman of the Joint Chiefs, who was on an official visit to China. It makes diplomatic relations more difficult and it shows that the military might not be as constrained by political oversight as we would want. We certainly are, so we think they are too. Every time there is confrontation with the West, the military is stronger. We need to tamp this down and do it in a way that is politically acceptable to both civilian governments.

Probably the best alternative is a "soft war" like the one we are having. I am not a politician and this is not a political solution, but we already had a cold war, and this does not feel the same. So we might as well recognize it as a different form of war. We are not at war with anyone, and we don't call it war. We need to learn to fight the information war on the scale that they are doing it. That is more difficult.

Those squeamish about even a cold war with China can say, "The Chinese are doing things that look like war to us." We have to believe that these moves of theirs are warlike. No matter what evidence there is, there will always be honest people, public relations firms, and a few governments that will disagree. We should listen, but carefully.

We need to look at what we buy from China and see if there is a way to limit their influence. In the last cold war, we traded with Russia, but we were careful about it and only traded for things we thought we couldn't get anywhere else. We just need to think of it the same way. If they want us to give them the capability to make the rope to hang us, we might want to keep that in mind and not give it to them. Once the federal bureaucracy cranks up, there will be no end to what we can do.

We can't do that with the government structure we have now. CFIUS is too slow to deal with the volume of companies trying to buy into our infrastructure, and a large part comes from overseas acquisitions where we don't have any influence. We needed some international cooperation here. Our industry leaders need to see this as the kind of threat it is, and to report any kind of attempt to buy into our systems, especially by China's state-owned companies and front companies. CFIUS is voluntarily reporting. The federal government has to deal with these seriously and quickly. We have to protect our telecommunications or we are going to get cut off one day. The Chinese are protecting theirs, so it should not be too hard for them to understand why we would want to.

If they are buying up the world's computer chips and telecommunications, then we need to start helping our businesses compete with that and shut off the sale of anything related to our national networks. When AT&T owned everything, we were better off, in some ways, than we are today. Somewhere along the way, we decided competition was good for the economy and would lower customer prices. It certainly did that, but we forgot about how important that base was to the country as a whole, and sold out our national security for consumer pricing. Our telecoms have to think too much about price and competition and not enough about national security. We need to give them some incentive to think about that more. We should not allow foreign competition—period—even with our friends.

There is quite a bit of spying in a soft war, and we need to increase ours—both the human kind and the kind with electronic gadgets of various sorts. This is the kind of spying done by the CIA and the rest of the intelligence community. With that goes counterintelligence, most of which is done by the FBI. We don't do enough of either one to even slow down what our adversaries are doing. We should be phasing down military operations to build up the CIA and FBI to handle these types of spying and counter-spying. It will take ten years to build up the forces that would be needed to counter the business and government spying that the Chinese are doing, so we don't need to be in a hurry, but we had better get started.

We need to learn from the Chinese. They understand information control and the effect it has on the world. They have done some smart things to control information and keep state secrets. We are far too open with some things and could benefit from their understanding, without building our own Golden Shield or intimidating our press corps. Sometimes we equate freedom with being able to say anything. With secrets, that can be harmful.

We will never stop the Russians, Chinese, Iranians or anyone else who wants to attack us until we have a deterrence strategy. Senator Lankford, a cyber-savvy member of Congress, has always asked hard questions because he understands the issues. At the Senate Intelligence Committee Global Threats hearing, in May 2017, he asked the leadership in the U.S. Intelligence Services about our strategy and if it was written down. He did not get a very good answer, and it was obvious that we do not have one written down. "It is coming," one respondent said, which was a weak excuse for inaction after years of discussion.

It is not a simple issue because it requires a capability to respond in kind to any attack that we label worthy of a response. There are two questions to be answered: What attack is worthy, and what response is appropriate. The Obama administration felt a cyber response to a cyber incident was not always required. But on the surface it seems that a response in kind is more effective—and more to the point—than response by other means. When the Trump administration launched cruise missiles after a chemical weapons attack in Syria, that was closer to a response in kind. We did not need to launch chemical warheads to make the point. It is obvious that those parts of a written policy are still under discussion. We have war plans on everything from the use of nuclear weapons in full-scale attacks to terrorist takeovers of friendly governments, but we cannot seem to get a strategy for dealing with cyber attacks.

Lankford followed with a question on what was an act of war in the cyber arena and got a better response. Director Pompeo said this was not something we should be discussing in an open session, something indisputably true. So far, the definition only includes attacks on the electrical grid, yet attacks on such things as the banking networks, air traffic control, health care networks and the federal payroll system would all require a response. These are the U.S. red lines. When an adversary crosses one of them, they have to know that they are going to get some retaliation, but that does not mean those red lines have to be publically defined. The Chinese and Russians make us guess on how far we can go before actions follow. The U.S. can use some of the same strategy.

Retaliation is a separate but related issue. When the U.S.–China Economic and Security Committee asked me about retaliation, suggesting that a more aggressive response might be in order, I told them that this kind of war has a way of escalating very quickly and the U.S. is not ready for the response. That reply offended some federal agencies that believe the U.S. is the world's leader in cyber. The military leadership, especially in Cyber Command, thinks it is ready and can do whatever is asked of it. That is not a very realistic assessment.

The example of Stuxnet, and Iran's response in attacking Aramco in Saudi Arabia, shows the dimensions. It is not limited to the country authorizing the actions or to the cyber combatants. The Chinese have been collecting information on U.S. leaders, business leaders, and government officials in anticipation of a day when they might need it. They have changed their strategy to penetrating systems and not taking information, making them more difficult to detect. They have gotten into government, business, and individual computer systems and they are still there. They are preparing for the kind of war that is going to be fought, not the kind the U.S. believes it can fight. And, for the first time in years, they have allies like Russia, Iran, and Syria who will help them. This speaks to the need for more U.S. resources in offensive information war.

Over the years, I have seen briefings that show the U.S. capabilities in offensive operations, but I tend to view them skeptically. Most of the time these briefings have confused our ability—or an adversary's ability—to collect intelligence with the ability to fight in cyberspace. Edward Snowden proved we were good at the former, but that does not translate into an offensive capability to attack, disrupt, or deny services on a scale that is required to deter these kinds of attacks. In fact, offensive operations often disrupt our own intelligence operations and need to be de-conflicted before they are carried out. Before we launch a retaliatory strike, we had better have the human and technical expertise to minimize the effects of an adversarial retaliation, and the unintended consequences of own actions. That includes trained people, sophisticated software, and an extremely secure environment to attack from. How much capability is needed is something that should be decided in places that can keep a secret and not in a public forum. In order to discuss it rationally, we must have realistic assessments of our own capabilities.

The second aspect of this is that our business community is not ready for any kind of war that retaliation would bring. There is very little the federal government can do about that, but plenty that the business leaders can do for themselves. The airlines, utilities and financial services industries need to have better coordination and a strategy to defend themselves against attacks. Movie studios may need it too, having been hacked by a group that wanted ransom to prevent the release of one of Disney's new movies. Business leaders used to do this regularly, but they appear to have no incentive to do it again.

The theft of information for strategic intelligence or economic benefit has become an important aspect that does not do damage, in a traditional sense of war, and in neither case is considered to be part of war. Yet the damage can accumulate to the point of providing strategic advantage. When Russian hackers got into data in the Democratic National Committee they could have well been looking for information of some intelligence value, but found something else. They may have initially kept quiet about that, saying nothing. There is no clear evidence that the Russians gave the information to Wikileaks and the founder, Julian Assange, said they did not.[2] Assange is certainly no friend to the United States, but somebody gave that information to him. But that does not mean he is not telling the truth about what happened. This kind of smoke and mirrors is part of a basic truth of information wars: no country tells the truth about what it does to other countries.

The dilemma we have about the credibility of Assange is the same as that we have about the U.S. Director of National Intelligence. Both want to protect their sources and methods—the means of collection—as state secrets in their own right. The collection by the National Security Agency of meta-data for billions of telephone calls actually does no war-like damage and gives the U.S. insight into networks of individuals who appear unconnected to any other person. Satellites from the National Reconnaissance Office show billions of images across the entire earth but they do no damage to anything. In both cases, the data allows a government to anticipate what will happen next, one of the major aspects of any strategic intelligence program. Some of those are things that another country does not want to have known, so they work to deny that information to others by putting their facilities underground, hiding the true purpose of what will be seen, or using the fact that they will be seen to their advantage. The collection of this kind of data is not war per se, but it can be used in preparation for war of a different kind.

Influence, or manipulation as I use the term here, is part of that war. In the 1930s, radio started to become a useful tool in getting out messages to the citizens of other countries. Russia controlled radio and centrally managed it. Germany beamed radio broadcasts into Russia, and encouraged its own citizens to buy radios. The French sent radio signals into Alsace, where they were jammed by the Germans. As time went on, more stations cropped up and were used to send messages to sympathetic ears.[3] But radio was also used as a clandestine device to communicate to agents operating in other countries. Several countries sent radio detection equipment out to find these stations and arrest the persons operating them. So what may appear to be just "playing music" may really be something else again, communicating with those sympathetic to the allied cause.[4] It could be for entertainment, which helps to keep a population more susceptible to other types of messages through the same medium. It can be a medium for directly engaging in war. But radio was simple in comparison to the media of today.

The Internet is a combination of other media such as computers, radio, telephones, and television. Marshall McLuhan said television was something bigger than the channels it carried; it was a complex medium. It integrates credible human beings with personas of many others. But the Internet is a super-medium that is far more powerful than any single medium like television. The Internet can be a tool for war that surpasses anything television or radio could ever be. We tend to think of the Internet as neutral, carrying anything we put on it, but that can be a deception. When governments use it to make messages that influence its citizens or citizens of another country, that can be part of a manipulation of ideas in a larger war. That is being done on scales we can barely perceive.

We are worried about Facebook offering a service like Facebook Live because people use it to show their own suicide or the rape of some total stranger. So Facebook says it will add people to try to find these offensive types of video and cut them off. That is censorship of sorts, but we accept that some of that has to be done. But there is something more important to us as citizens than Facebook Live.

There are too many Facebook friends and Linkedin business associates who are not real people. They are personas, just like the personas used in the Hillary Clinton e-mail scandal. They tout a party line of another country as if some civic-minded citizen of the world wants to comment on world events or policies of other governments. But they are paid to do something else—targeting specific groups to recruit, to target with phishing schemes, or to influence. Iran used Linkedin to create fake profiles indicating the users were high-profile business leaders to do exactly that.[5] It is almost impossible for service providers to find these fake accounts because they are kept current by the Iranians operating them as a normal user would.

In the same way, political parties use paid "media assistants" to promote candidates and sell political ideas. They may use their real names, or multiple personas, but their motivation is not the same as a normal user. What we found in the dissemination of news stories and propaganda was that world intelligence services were using these kinds of accounts to spread their versions of events. Mark Zuckerberg, who owns Facebook, spoke about this in April 2017, indicating that information operations activities go well beyond the "fake news" that is being disseminated.[6] He said this is about the collection of information on specific individuals and stealing passwords, among other things. Facebook has decided to use specialized software to look for these kinds of accounts and remove them. That may prove more difficult than they think. Facebook is big, but they are only the tip of the iceberg for this kind of activity.

Our private businesses need to know when users are not who they say they are. About the only organizations that can tell them that are the intelligence services, which are not tasked to do any such thing. What they can be tasked to do is find out when these phony accounts are being created by intelligence services, and have a mechanism to identify those accounts to the service providers. It will be a lifetime job for anyone doing it, but of value to all the users of these systems.

The question of course is how far do we go in trying to find fake news and users who are not who they say they are? It is easy to step over the line, from looking for authentic users to the extremes of censorship we see in China, and in development in many other countries. We have users on the left of our political spectrum who believe free speech is only free as long as it is consistent with their political views, and they have many friends. They would be ready to man those censorship ports looking for ideas that would not fit their own views, and happy to persecute people for the greater good.

Some countries have decided the Internet is too dangerous to be used by its citizens without supervision. China, Russia, Syria, Egypt, Saudi Arabia, and the United Arab Emirates are just a few of the countries that believe their citizens need to be watched while they are on the Internet. They use sophisticated tools to monitor cell phones, computers, web sites and e-mail and take action against people who use those in ways inconsistent with government policy.

In democracies, private companies and government offices use those same tools in monitoring employees for policy violations. The U.S. might not see itself as a country that monitors its citizens, but there is much more monitoring going on than most of its citizens know. Amazon, Facebook, and Google know more about our citizens than almost any other services, and they sometimes sell information about us. These services know where we are, what websites we visit, and when we call for an Uber pickup. We pretend that is a cost of using their services. But if we do not like what they are doing to preserve our privacy, the only recourse is to vote by changing services, which will not solve the problem.

It is no coincidence that these issues are spreading in places that encourage free speech and are constrained in places that do not. On both sides, this is a battle between governments that see themselves as representatives of the people they serve. In one, the government has to do everything to help form the ideas of the people to help them be in harmony with the state and each other. The leaders of those countries stay in power by some of the information war concepts described in this book. On the other side, the government is a service to the citizens who have a right to elect new leadership or sponsor new ideas that may not fit well with the plans of the central government. Those countries are unruly, disjointed, and entertaining, but they are losing this war.

It is not the people of those countries that make war. Governments make war. They do it today in secret, and they do it without the consent of their citizens. They control narratives and the press, and they manage people who have a different opinion. Their armed forces back up their actions. They seize territory and justify it to the world using the same techniques. We can't fight back without a better understanding of how information warfare actually works, and the development of techniques to counter the strategies being employed.

There is much more to information war than cyber. In the economic sector we have conceded too much to the Chinese ability to manufacture goods. The world calls that globalization without mentioning the inequities of trade by countries that manipulate normal business relationships through mechanisms that are biased towards domestic production. Changing ownership requirements, banning goods based on bureaucratic whim, and adding taxes at the border are things that can be undone through reciprocal policies that point out how inequitable world trade already is.

The greatest advance in settling these kinds of inequities has shown up in the Trump administration: the use of the word *reciprocity*. This is a simple-sounding word that most people believe they understand, but most people are not economists. It seems as though reciprocity should mean that when China puts a 10 percent tariff on automobiles coming in, each country selling them autos would put a 10 percent tariff on their autos coming out. That is what the Trump administration is saying. But that is not what economists say it means.[7]

Economists understand the Reciprocal Trade Agreements Act of 1934, which says if China puts a 10 percent tariff on autos, each country has to put a 10 percent tariff on every other country's autos to match China's. There is a concept of Most Favored Nation, which encourages this kind of behavior, and indicates why economists should not be running the trade agreements between countries. The World Trade Organization tries to keep all of this straight, but is too slow to deal with China, which takes advantage of their inability to keep up. It takes months to get a complaint together, and more months to get one heard. By the time the WTO gets a ruling, China concedes and moves on. Reciprocity implies that we can take similar action towards China without filing formal complaints that take too long to resolve.

We also need a similar understanding of foreign ownership, control, and influence that is the equivalent of a policy for reciprocity. CFIUS is too slow to monitor and bring action against the number of companies trying to buy the U.S. infrastructure. The Chinese adapt quickly, and within a year of taking on Huawei, replaced it as the purchaser of technology components in places CFIUS had oversight. They diversified and diffused the buyers, making it more difficult to decide when a central purchase was being made by the Chinese government. This brought to the fore the issue of whether a purchase by a government-owned entity is a purchase by the government of China.

With the number of countries having government-owned enterprises, it seems that this issue should have been resolved long ago.

In addition, the Chinese are too casual about allowing violations of sanctions they vote for in collaborative bodies like the United Nations, and they are not accountable for their actions. Oversight of that type of action requires more action by the intelligence community in monitoring compliance with sanctions. North Korea and Iran have advanced their causes because China is unwilling to enforce sanctions they have agreed to. That requires exposure of performance on sanctions, which the Obama administration did when it sanctioned ZTE. But while the idea of exposing them was novel, there has to be much more of this to make enforcement possible. If these sanctions are being violated every day, we should be seeing some action every week to bring them to the attention of the enforcement agencies.

Even more important, we need a better understanding of the kind of war China is fighting. Annexation is not going to give us that. All the use of that term does is prolong the time for fighting of the type that is required to stop their behavior. China threatens the U.S. both blatantly and through its proxies. Ignoring those threats is allowing them to push that far beyond what should be acceptable to another country. For the first time in a decade, China is being held responsible for the trouble it causes through North Korea. That step has already shown some benefit, but it is a long journey only just begun.

Our foreign policy leaders have to say out loud that China cannot have Taiwan, the South China Sea, or the trade routes that go with them. Every time they take another step in that direction, there has to be some response by the U.S. and its allies. The diplomatic approach seems to turn away.

China is ruthless at meeting its political agenda with the technical means necessary. Managing information people see and hear is a clever but expensive process. They have tried to simplify it. The April 2015 announcement by Google that it was no longer accepting digital certificates in Chrome from the Chinese Network Information Center (CNIC) is a good indicator. Digital certificates are the basis for a user to know when they are communicating with a service that is legitimate. The lock that appears in the browser window of a Chrome browser gives a user some assurance that the site is who it says it is. Because this is an important aspect to both vendors and users, there are controls placed on who can issue a certificate and under what conditions. If everyone follows the rules, we won't have a site calling itself Amazon.com collecting credit card numbers for the real Amazon.com, or looking at what forbidden items are being purchased. The citizens of China know this much better than other citizens of the world.

In a *New York Times* op-ed, Murong Xeucun, an anti-censorship advocate who spent three years in Lhasa, Tibet, was asked by a friend if he knew about people setting themselves on fire in protest to Chinese governance. He hadn't. His friend said, "Everyone beyond the wall knows this. A writer who cares about China, but who doesn't go over the wall, suffers from a moral deficiency. You shouldn't let a wall decide what you know."[8] The wall he was talking about is the Great Firewall, and the Chinese people have found a few ways around it. When the U.S. considered sanctions against China for cyber theft, it considered improving those ways, and should have continued down that path. It might be useful to consider the same kind of response to Russia. In other words, punish these countries by circumventing their ability to control what is being said to their own populations. It was a good idea, apparently never implemented. The Chinese have to do something on their own to get around it.

It is a risky game being played with virtual private networks (VPNs), which the government searches for. Murong said he tried using VPNs but his first one was detected after three months. He got a different one, which went on for a time; and he got a succession of them over the years. Some people in his account had their Internet accounts terminated and some were arrested. In each case these individuals say they are "walled." China wages this battle with its own people and does not let up.

Gary King, at Harvard, studied how effective the censorship of information in China really is. King systematically studied the process by starting a web site in China and using software intended for use there. He found that censorship was not exactly what we think it is. It was a three-step process of automated review, set-asides for questionable material, and human review for the final determination. Of things that go to human review, 63 percent never get to the web.[9]

The review process does not censor criticism of the government unless it is connected to what is called a "real world collective action event," described as "those which (a) involve protest or organized crowd formation outside the Internet; (b) relate to individuals who have organized or incited collective action in the past; or (c) related to nationalism or nationalistic sentiment that have incited collective action in the past."[10] It was carried out by the web sites themselves, who had software and rules to follow in the administration. They had wide flexibility in how to apply the rules, but the rules were fuzzy enough to have things censored that were not necessary. It also censored things that might praise the government but also related to a collective action event. The major factor was the effectiveness of reviewing material first, then publishing it to the web. It is neither entirely manual, nor subject to the randomness of human reviewers, but it was effective. Even so, this part of censorship manages information given to users of media in China.

This approach is a step beyond blocking and filtering done on a firewall, launching attacks on sites where the Chinese want to discourage access. Baidu denied being involved in the denial of service attack, but their ad software seemed to be the source. Baidu's browser technology came under scrutiny by Citizen Lab because it acquired and sent user search terms, hard drive serial number, and much more that was unrelated to their business. These are all things that have little legitimate use to anyone operating a company doing network services, but have a great deal of intelligence value. As in the case of the Great Firewall, we would conclude government involvement across companies in China that, by comparison to the Edward Snowden disclosures of U.S. capabilities, makes it look like China has an equal capability. That is not the only aspect that makes the Chinese more dangerous than other countries. In times of crisis, they have the ability to manage what is on the Internet and what people in their country do with it. Though the Russians want to try the same techniques, like the rest of the world, they are far behind.

The Chinese have one governmental trait that we have to admire: they set out on a path with a long-term goal, and they continue on that path, sweeping around obstacles, managing the messages to fit their view of an issue, and doing what is needed to complete tasks required to meet their objectives. They pay attention to what other governments say about their actions, and they respond to mitigate negative feelings. As much as possible, they stay under the radar to avoid attracting attention, but they are persistent and ruthless at doing what they say they are going to do.

Four years ago, there was not much evidence to support a contention that the Chinese were stealing information from businesses and plowing it back into their economy. They denied it, and the number of cases detected didn't give us much cause for alarm. That part is considerably different today. The range of things stolen by China now extends from weapon systems designs of sensitive government programs to seed corn in the fields of Iowa. Now that we look for Chinese thefts, we find more of them, but we still have not identified all of the things they have taken, nor how long some of those thefts have gone undetected.

Four years ago, there were few authoritative sources saying cyber thefts were executed on the orders of the Chinese government, and China still denies that the effort shown to steal trade secrets is based on a national strategy to improve their commercial products. The Chinese deny anything attributed to them, yet accusations now follow them everywhere.

Since an agreement between the U.S. and China in September 2015 to refrain from "conducting or knowingly supporting cyber-enabled theft of intellectual property with the intent of providing competitive advantage to companies or commercial sectors," the U.S. Director of National Intelligence noted that commercial businesses "have identified limited ongoing cyber activity from China but have not verified state sponsorship or the use of exfiltrated data for commercial gain."[11] This politically correct statement is not the whole story, since industrial sources are not responsible for coming to conclusions about what the Chinese government directs. The U.S. intelligence community, which the director heads, is responsible for drawing those conclusions. This statement allowed the Obama administration to claim that China was in compliance with the understanding reached with them in September 2015, because there "is no evidence they have not been complying." While the Chinese violated the sanctions imposed on Iran and North Korea, we somehow believed they would keep agreements. The following month, a *Washington Post* editorial claimed the Chinese continue to steal trade secrets and have made little effort to cease those programs already underway.[12]

We need to disrupt the steady progress our enemies are making. But before we can start, we have to understand that this is more than just normal interaction between countries. We are too quick to dismiss an ulterior motive for some of their actions and accept the denials of government officials with terrible track records for the accuracy of their statements. The soft war between us is more substantive than we are willing to accept, yet territorial losses speak to the content of an information war. It is becoming harder and harder to make excuses for actions that seize territory or blatantly take technology to compete on unequal grounds.

There are a number of democracies in the world that tend to look at problems with China as solely those of the United States. They see the conflict as the number one and number two economies in the world battling in a free trade exercise of competition. They are missing a good bit of the conflict. The underlying clashes are generating pressure that is intended to disrupt democratic institutions and overrule the wishes of the people who elect their own leadership. We owe each other a harder look at what can be done to stop them.

Chapter Notes

Chapter 1

1. Henry Campbell Black et al., *Black's Law Dictionary* (St. Paul, MN: West, 1990), p. 88.
2. Ibid., p. 1583.
3. Mao Zedong. *Mao Tse-Tung on Protracted War* (Peking: Foreign Languages Press, 1967).
4. See http://www.pcr.uu.se/research/ucdp/faq/ for various areas of research and documents.
5. Erik Melander, *Encyclopedia of Political Thought* (n.p.: John Wiley, 2014), extract from http://uu.diva-portal.org/smash/record.jsf?pid=diva2%3A799013&dswid=-335.
6. The British Broadcasting System, "Migrant Crisis: Migration to Europe Explained in Seven Charts," 4 March 2016, http://www.bbc.com/news/world-europe-34131911.
7. Winston Churchill, "The Churchill Society, Churchill's Wartime Speeches," excerpt from *The Munich Agreement, A Total and Unmitigated Defeat*, House of Commons, 5 October 1938, http://www.churchill-society-london.org.uk/Munich.html.
8. Elias Groll, "'Obama's General' Pleads Guilty to Leaking Stuxnet Operation," *Foreign Policy*, 17 October 2016, http://foreignpolicy.com/2016/10/17/obamas-general-pleads-guilty-to-leaking-stuxnet-operation/.
9. Fred Kaplan, *Dark Territory* (New York: Simon and Shuster, 2016).
10. CIA, *Remarks as Prepared for Delivery by Central Intelligence Agency Director Mike Pompeo at the Center for Strategic and International Studies*, 13 April 2017, https://www.cia.gov/news-information/speeches-testimony/2017-speeches-testimony/pompeo-delivers-remarks-at-csis.html.
11. Christian Caryl, *Novorossiya Is Back from the Dead*, Foreign Policy, 17 April 2014, http://foreignpolicy.com/2014/04/17/novorossiya-is-back-from-the-dead/.
12. Department of Defense, *Annual Report to Congress, Military and Security Developments, Involving the People's Republic of China 2017*, p. 45.

Chapter 2

1. George J. Tenet, "DCI Testimony Before the Senate Select Committee on Government Affairs," 24 June 1998, https://www.cia.gov/news-information/speeches-testimony/1998/dci_testimony_062498.html.
2. Roger C. Molander, Andrew S. Riddile, and Peter A. Wilson, *Strategic Information Warfare* (Santa Monica, CA: Rand, 1996), p. 1.
3. Zalmay Khalilzad and John P. White, *The Changing Role of Information in Warfare* (Santa Monica, CA: Rand, 1999).
4. Wang Baocun (senior colonel) and Li Fei, "Information Warfare," summarized from articles in the *Liberation Army Daily*, 1995.
5. Paul Mozur and Janie Perlez, "China Bets on Sensitive U.S. Start-Ups, Worrying the Pentagon," *The New York Times*, 22 March 2017, https://www.nytimes.com/2017/03/22/technology/china-defense-start-ups.html.
6. Munk Centre for International Studies, Citizen Lab, Shadowserver Foundation, and Information Warfare Monitor, *Shadows in the Cloud: Investigating Cyber Espionage 2.0* ([Toronto, Ont.]: [Citizen Lab, Munk Centre for International Studies, University of Toronto], 2010).
7. James R. Clapper, Director of National Intelligence, before the House Permanent Select Committee on Intelligence, 10 September 2015.
8. Ellen Nakashima, *When Is a Cyberattack an Act of War? The Washington Post*, 26 October 2012, https://www.washingtonpost.com/opinions/when-is-a-cyberattack-an-act-of-war/2012/10/26/02226232-1eb8-11e2-9746-908f727990d8_story.html.
9. Garance Burke and Jonathan Fahey, *The Times of Israel*, 22 December 2015, http://www.

timesofisrael.com/iranian-hackers-breached-us-power-grid-to-engineer-blackouts/.

10. Kim Zetter, "U.S. Considered Hacking Libya's Air Defense to Disable Radar," *Wired*, 17 October 2011, http://www.wired.com/threatlevel/2011/10/us-considered-hacking-libya/.

11. Krebs on Security, "Who Else Was Hit by the RSA Attackers," http://krebsonsecurity.com/2011/10/who-else-was-hit-by-the-rsa-attackers/.

12. "China's First Aircraft Carrier 'starts first sea trials,'" BBC News, 10 August 2011, http://www.bbc.co.uk/news/world-asia-pacific-14470882.

13. Eli Lake, "China Bid Blocked over Spy Worry," *Daily Beast*, 11 October 2011.

14. Juro Osawa and Eva Dou, "U.S. to Place Trade Restrictions on China's ZTE," *The Wall Street Journal*, 7 March 2016.

15. Department of Commerce, Bureau of Industry and Security, "Proposal for Import and Export Control Risk Avoidance," internal document of ZTE posted in English at https://www.bis.doc.gov/index.php/forms-documents/about-bis/newsroom/1436-proposal-for-english/file.

16. Juro Asawa, "ZTE to Replace Three Senior Executives," *The Wall Street Journal*, 2 April 2016.

17. Eva Dau, "China to Start Security Checks on Technology Companies in June," *The Wall Street Journal*, 3 May 2017, https://www.wsj.com/articles/china-to-start-security-checks-on-technology-companies-in-june-1493799352.

18. U.S.–China Economic and Security Review Commission, *2010 Annual Report to Congress*, 111th Congress, 2nd Session, November 2010, p. 244.

19. Select Committee of the U.S. House of Representatives, "U.S. National Security and Military/Commercial Concerns with the People's Republic of China," June 2005.

20. BBC News, World Americas, "China spying for 20 years," 26 May 1999, http://news.bbc.co.uk/2/hi/americas/352591.stm

21. *The New York Times*, "Nuclear Secrets: What China knows about U.S. Missile Technology, A Chronology, 26 May 1999," http://www.nytimes.com/1999/05/26/world/nuclear-secrets-what-china-knows-about-us-missile-technology-a-chronology.html

22. U.S.–China Economic and Security Review Commission, *2016 Annual Report to Congress*, p. 157.

23. David E. Sanger and Nicole Perlroth, *N.S.A. Breached Chinese Servers Seen as Security Threat*, The New York Times, 22 March 2014, https://www.nytimes.com/2014/03/23/world/asia/nsa-breached-chinese-servers-seen-as-spy-peril.html.

24. Department of Defense, "Joint Publication 3–13, Information Operations," 20 November 2014, p. I-4.

25. Brian C. Lewis, "Information Warfare," original source not identified, https://fas.org/irp/eprint/snyder/infowarfare.htm.

26. Ibid.

27. See CIA, "What We Do," https://www.cia.gov/about-cia/todays-cia/what-we-do, for this mission, which is not explained.

28. Clay Wilson, "Information Operations, Electronic Warfare, and Cyberwar: Capabilities and Related Policy Issues," Congressional Research Service, 20 March 2007.

29. Ibid., citing John Lasker, "U.S. Military's Elite Hacker Crew," *Wired News*, April 18, 2005, http://www.wired.com/news/privacy/0,1848,67223,00.html; U.S. Strategic Command Fact File, http://www.stratcom.mil/fact_sheets/fact_jtf_gno.html; U.S. Strategic Command Fact File, http://www.stratcom.mil/fact_sheets/fact_jioc.html.

30. "Agence France-Presse in Berlin, Russia Accused of Series of International Cyber-Attacks," *The Guardian*, 13 May 2016.

31. Clapper, James R. Verbal testimony. Emerging Threats Before the U.S. House of Representatives Committee on Homeland Security Subcommittee on Cybersecurity, Infrastructure Protection, and Security Technologies, 19 November 2016.

Chapter 3

1. Sun Tzu, *The Art of War* (London: Luzac, 1910).

2. Ben Blanchard and Benjamin Kang Lim, "'Give Them a Bloody Nose': Xi Pressed for Stronger South China Sea Response," Reuters, 31 July 2016, http://www.reuters.com/article/us-southchinasea-ruling-china-insight-idUSKCN10B10G.

3. Iaconangelo, David, "Why are Chinese protesters picketing KFC and smashing iPhones?," Christian Science Monitor, 20 July 2016, https://www.csmonitor.com/World/Asia-Pacific/2016/0720/Why-are-Chinese-protesters-picketing-KFC-and-smashing-iPhones.

4. Lincoln Feast and Greg Torode, "Exclusive: Risking Beijing's ire, Vietnam Begins Dredging on South China Sea Reef," Reuters, 9 December 2016, http://www.reuters.com/article/us-south china-sea-vietnam-idUSKBN13X0WD.

5. Jesse Johnson, "China Deploys Anti-Diver Rocket Launchers to Man-Made Island in South China Sea: Report," *The Japan Times*, undated article, http://www.japantimes.co.jp/

news/2017/05/17/asia-pacific/china-deploys-anti-diver-rocket-launchers-man-made-island-south-china-sea-report/#.WR8i4jOZOL8.

6. Andrew Browne, "Man in the Middle: Rodrigo Duterte Gets a Taste of China's Heavy Hand," *The Wall Street Journal*, 19 July 2016, http://www.wsj.com/articles/man-in-the-middle-rodrigo-duterte-gets-a-taste-of-china-1468909553.

7. Zhao Lei, "New Satellite Keeps Eye on Sea Interests," *China Daily*, 11 August 2016, http://www.chinadaily.com.cn/china/2016–08/11/content_26426328.htm.

8. Sean O'Conner, "Imagery Shows Chinese HQ-9 Battery Being Removed from Woolly Island." Jane's 360, 21 July 2016, http://www.janes.com/article/62442/imagery-shows-chinese-hq-9-battery-being-removed-from-woody-island.

9. Federation of American Scientists, "2018 Nuclear Posture Review Resource," http://fas.org/nuke/guide/dprk/missile/td-2.htm.

10. Travis Wheeler, "China's MIRVs: Separating Fact from Fiction," *The Diplomat*, 18 May 2017, http://thediplomat.com/2016/05/chinas-mirvs-separating-fact-from-fiction/.

11. Arms Control Association, "U.S. Missile Defense Programs at a Glance," August 2016, https://www.armscontrol.org/factsheets/us missiledefense.

12. Andrew Roth, "Russia and China Sign Cooperation Pacts," *The New York Times*, 8 May 2015, https://www.nytimes.com/2015/05/09/world/europe/russia-and-china-sign-cooperation-pacts.html.

13. Associated Press, "Lithuanian 'Elves' Combat Russian Influence On Line," 28 December 2016, http://www.foxbusiness.com/features/2016/12/28/lithuanian-elves-combat-russian-influence-online.html.

14. Charlotte McDonald-Gibson, "Europe Mulls a Russian Language TV Channel to Counter Moscow Propaganda," Time, 19 January 2015, http://time.com/3673548/europe-russian-language-tv/.

15. Maria Tsvetkova, "Special Report: Russian Fighters, Caught in Ukraine, Cast Adrift by Moscow," Reuters, 29 May 2015, http://www.reuters.com/article/us-ukraine-crisis-captured-specialreport-idUSKBN0OE0YE20150529.

16. David Francis, "U.S. Treasury Hits Russia with More Sanctions over Ukraine." *Foreign Policy*, 1 September 2016, http://foreignpolicy.com/2016/09/01/u-s-treasury-hits-russia-with-more-sanctions-over-ukraine/.

17. See BBC, "MH17 Ukraine Plane Crash: What We Know," 28 September 2016, http://www.bbc.com/news/world-europe-28357880.

18. Russia Today, "Crimean Energy Bridge Completed from Mainland Russia," 12 May 2016, https://www.rt.com/business/342737-crimea-energy-bridge-complete/.

19. Russia RT, "Emails Expose Watchdog's Dollar Deals," 8 December 2011, https://www.rt.com/news/election-america-golos-support-393/.

20. Levin, Dov, "Sure, the U.S. and Russia often meddle in foreign elections. Does it Matter?" *The Washington Post*, 7 September 2016.

21. Art Swift, "Americans' Trust in Mass Media Sinks to New Low," Gallup, 14 September 2016, http://www.gallup.com/poll/195542/americans-trust-mass-media-sinks-new-low.aspx; and Damien Sharkov, "Half of Russians Do Not Trust Russian Media: Poll," *Newsweek*, 17 October 2016, http://www.newsweek.com/majority-russians-do-not-trust-national-media-510682.

22. Amy Mitchell et al., "Trust and Accuracy from the Modern News Consumer," Pew Research Center, 7 July 2016, http://www.journalism.org/2016/07/07/trust-and-accuracy/.

23. Elen Aghekyan, Bret Nelson, et al., *Freedom of the Press 2016* (Washington, D.C.: Freedom House, 2016).

24. Video: BBC, "Kung Fu Grandma Is China's New Internet Sensation," http://www.bbc.com/news/video_and_audio/headlines/39073334.

25. Konstantin Benyumov, "How Russia's Independent Media Was Dismantled Piece by Piece," *The Guardian*, 25 May 2016, https://www.theguardian.com/world/2016/may/25/how-russia-independent-media-was-dismantled-piece-by-piece.

Chapter 4

1. George Morgenstern, "The Actual Road to Pearl Harbor," in *Perpetual War for Perpetual Peace*, ed. Harry Elmer Barnes, pp. 332–343 (Caldwell, Idaho: Caxton, 1953).

2. Paul A. Smith, *On Political War* (Washington, D.C.: National Defense University Press, 1989), pp. 172–173.

3. Max Boot, Jeane J. Kirkpatrick, et al., "Political Warfare," Council on Foreign Relations, June 2013, http://www.cfr.org/wars-and-warfare/political-warfare/p30894.

4. CIA, "Comments on [redacted] Study 'The Vulnerability of the Soviet Union and Its European Satellites to Political Warfare,'" Central Intelligence Agency Redacted Report, declassified 5 June 2013, https://www.cia.gov/library/readingroom/docs/CIA-RDP61S00750A000300100055–6.pdf.

5. Ellen Nakashima, "When Is a Cyberattack an Act of War?" *The Washington Post*, 26 Octo-

ber 2012, https://www.washingtonpost.com/opinions/when-is-a-cyberattack-an-act-of-war/2012/10/26/02226232–1eb8–11e2–9746–908 f727990d8_story.html.

6. Evan Perez, "U.S. Official Blames Russia for Power Grid Attack in Ukraine," Cable News Network, 11 February 2016, http://www.cnn.com/2016/02/11/politics/ukraine-power-grid-attack-russia-us/.

7. Kim Zetter, "Inside the Cunning Unprecedented Hack of Ukraine's Power Grid," *Wired*, 3 March 2016, https://www.wired.com/2016/03/inside-cunning-unprecedented-hack-ukraines-power-grid/.

8. Bruce Klinger, "Chinese Foot-Dragging on North Korea Thwarts U.S. Security Interests," The Heritage Foundation, 11 April 2016, http://www.heritage.org/research/reports/2016/08/chinese-foot-dragging-on-north-korea-thwarts-us-security-interests.

9. Nikko Dizon and Nina P. Callega, "PH: China '9-Dash Line' Doesn't Exist," *Philippine Daily Inquirer*, 24 November 2015.

10. Stacy Hsu, "Presidential Office Rejects Criticism," *Taipei Times*, 11 May 2016, http://www.taipeitimes.com/News/taiwan/archives/2016/05/11/2003646004.

11. Chun Han Wong, "China Appears to Have Built Radar Facilities on Disputed South China Sea Islands," *The Wall Street Journal*, 23 February 2016.

12. See Asia Maritime Transparency Initiative, "Airpower Projection," http://amti.csis.org

13. *The Guardian,* "Agence France-Presse, South China Sea: Beijing Tells G7 Foreign Ministers to Keep Out of Territorial Dispute," 12 April 2016.

14. Shannon Tiezzi, "China Push for an Asia-Pacific Free Trade Agreement," *The Diplomat*, 30 October 2014.

15. Prashanth Parameswaran, "China Enforcing Quasi-ADIZ in South China Sea: Philippine Justice." *The Diplomat*, 13 October 2015.

16. Sam LaGrone, "PACOM Harris: U.S. Would Ignore a 'Destabilizing' Chinese South China Sea Air Defense Identification Zone," *U.S. Naval Institute News*, 26 February 2016.

17. Ronald O'Rourke, "Maritime Territorial and Exclusive Economic Zone (EEZ) Disputes Involving China: Issues for Congress," Congressional Research Service, United States Congress, 22 December 2015.

18. The U.S.–China Economic and Security Review Commission 2015 Report to Congress.

19. Josh Chin, "Cyber Sleuths Track Hacker to China's Military," *The Wall Street Journal*, 23 September 2015.

Chapter 5

1. Sam Thielman and Spencer Ackerman, "Cozy Bear and Fancy Bear: Did Russians Hack Democratic Party and If So, Why?" *The Guardian*, 16 July 2016, https://www.theguardian.com/technology/2016/jul/29/cozy-bear-fancy-bear-russia-hack-dnc.

2. Ned Parker, Jonathan Landay and John Walcott, "Putin-Linked Think Tank Drew Up Plan to Sway 2016 Election—Documents," Reuters, 21 April 2017, http://www.reuters.com/article/us-usa-russia-election-exclusive-idUSKBN17L2N3.

3. Threat Intelligence, "APT28: A Window into Russia's Cyber Espionage Operations?" 27 October 2014, https://www.fireeye.com/blog/threat-research/2014/10/apt28-a-window-into-russias-cyber-espionage-operations.html.

4. Ibid.

5. FireEye Threat Intelligence, "Hammertoss: Stealthy Tactics Define a Russian Cyber Threat Group," 29 July 2015, https://www.fireeye.com/blog/threat research/2015/07/hammertoss_stealthy.html.

6. Toomas Hendrik Ilves, "Prepared Testimony: Undermining Democratic Institutions and Splintering NATO: Russian Disinformation," The House Foreign Affairs Committee, March 9, 2017, http://docs.house.gov/meetings/FA/FA00/20170309/105674/HHRG-115-FA00-Wstate-IlvesH-20170309.pdf.

7. Carl Schreck, "Russian Lawyer Says FSB Officers, Kaspersky Manager Charged with Treason," Radio Free Europe/Radio Liberty, 1 February 2017, http://www.rferl.org/a/russia-fsb-officers-treason-kaspersky/28272937.html.

8. David E. Sanger and Nicole Perlroth, "As Democrats Gather, a Russian Subplot Raises Intrigue," *The New York Times*, 24 July 2016, https://www.nytimes.com/2016/07/25/us/politics/donald-trump-russia-emails.html.

9. The U.S. Justice Department says at least 500 million accounts were taken relating to the charges filed; U.S. Justice Department, "U.S. Charges Russian FSB Officers and Their Criminal Conspirators for Hacking Yahoo and Millions of Email Accounts," press release, 15 March 2017, https://www.justice.gov/usao-ndca/pr/us-charges-russian-fsb-officers-and-their-criminal-conspirators-hacking-yahoo-and.

10. British Broadcasting Service, "Democrat Hack: Who Is Guccifer 2.0?" 28 July 2016.

11. Jonathan Martin and Alan Rappeport, "Debbie Wasserman Schultz to Resign D.N.C. Post," *The New York Times*, 24 July 2016, https://www.nytimes.com/2016/07/25/us/politics/

debbie-wasserman-schultz-dnc-wikileaks-emails.html?_r=0.

12. Cory Bennett, "Guccifer 2.0 Drops More DNC Docs," *Politico*, 13 September 2016.

13. Michael Sainato, "Exclusive: Wikileaks Guccifer 2.0 Teaser Exposes Pay-to-Play and Financial Data," *The New York Observer*, 5 October 2016, http://observer.com/2016/10/exclusive-wikileaks-guccifer-2-0-teaser-exposes-pay-to-play-and-financial-data/.

14. Katie Bo Williams, "NSA Head: DNC Hack Didn't Affect Election Outcome," *The Hill*, 21 November 2016, http://thehill.com/policy/cybersecurity/307031-nsa-head-dnc-hack-didnt-impact-election-outcome.

15. Laura Vozzella and Simon Denyer, "Donor to Clinton Foundation, McAuliffe Caught Up in Chinese Cash-for-Votes Scandal," *The Washington Post*, 16 September 2016, https://www.washingtonpost.com/local/virginia-politics/clinton-foundation-mcauliffe-donor-caught-up-in-chinese-cash-for-votes-scandal/2016/09/16/bfb3b8fc-7c13-11e6-ac8e-cf8e0dd91dc7_story.html?utm_term=.47cf8561f1b7.

16. Nicholas Confessore and Stephanie Saul, "Inquiry Highlights Terry McAuliffe's Ties to Chinese Company," *The New York Times*, 24 May 2016, https://www.nytimes.com/2016/05/25/us/politics/terry-mcauliffe-wang-wenliang.html.

17. John Roth, "Investigation into Employee Complaints About the Management of U.S. Citizenship and Immigration Services' EB-5 Program," Inspector General Homeland Security, 24 March 2015, https://www.oig.dhs.gov/assets/Mga/OIG_mga-032415.pdf.

18. Nicholas Colucci, "Testimony Before the Senate Judiciary Committee," 2 February 2016, https://www.judiciary.senate.gov/imo/media/doc/02-02-16%20Colucci%20Testimony.pdf. Note: Colucci added, in response to a follow-up clarification question, that the agency did not have sufficient data to show the location of all the projects funded under this program.

19. Alana Semuels, "Should Congress Let Wealthy Foreigners Buy Green Cards?" *The Atlantic*, 21 September 2015, https://www.theatlantic.com/business/archive/2015/09/should-congress-let-wealthy-foreigners-buy-citizenship/406432/.

20. CCTV Global Business, "China's Dandong Port Co. Could Benefit from Democratic National Convention," 7 September 2012, https://youtube.com/watch?v=OiZm2hIMDs4&feature=youtu.be.

21. Sophia Yan, "U.S. Runs Out of Investor Visas Again as Chinese Flood Program," CNN, 15 April 2015, http://money.cnn.com/2015/04/15/news/economy/china-us-visa-eb5-immigrant-investor/.

22. David E. Sanger, "Pentagon Announces New Strategy for Cyberwarfare," *The New York Times*, 23 April 2015.

23. Steward Baker, "NSA Files," *The Guardian*, accessed November 2015, http://www.theguardian.com/world/interactive/2013/nov/01/snowden-nsa-files-surveillance-revelations-decoded#section/7.

24. See Ron Deibert, "Cyber Crime and Warfare," University of Toronto, accessed November 2015, https://www.youtube.com/embed/PqvKGfDXFE4?list=UUu_u-P3cBFO7D-sAjxd_I-w.

25. Ibid.

26. San-Hun Choe, "Computer Networks in South Korea Are Paralyzed in Cyberattacks," *The New York Times*, 20 March 2013.

27. Kara Scannell, "FBI Details North Korean Attack on Sony," *The Financial Times*, 7 January 2015, https://www.ft.com/content/287beee4-96a2-11e4-a83c-00144feabdc0.

28. David E. Sanger and Martin Fackler, "N.S.A. Breached North Korean Networks Before Sony Attack, Officials Say," *The New York Times*, 18 January 2015.

29. Kim Zetter, "Experts Are Still Divided on Whether North Korea Is Behind Sony Attack," *Wired*, 23 December 2014.

30. Chris Strohm, "North Korea Web Outage Response to Sony Hack, Lawmaker Says," Bloomberg L.P., 17 March 2015, accessed 26 October 2015, http://www.bloomberg.com/politics/articles/2015-03-17/north-korea-web-outage-was-response-to-sony-hack-lawmaker-says; Carol E. Lee and Jay Solomon, "U.S. Targets North Korea in Retaliation for Sony Hack," *The Wall Street Journal*, 3 January 2015.

31. Ellen Nakashima, "U.S. Developing Sanctions Against China over Cyberthefts," *The Washington Post*, 30 August 2015, https://www.washingtonpost.com/world/national-security/administration-developing-sanctions-against-china-over-cyberespionage/2015/08/30/9b2910aa-480b-11e5-8ab4-c73967a143d3_story.html?utm_term=.d020b3917047.

32. Eduard Kovacs, "APT3 Hackers Linked to Chinese Ministry of State Security," *Securityweek*, 17 May 2017, http://www.securityweek.com/apt3-hackers-linked-chinese-ministry-state-security.

33. Ellen Nakashima, "Following U.S. Indictments, China Shifts Commercial Hacking away from Military to Civilian Agency," *The Washington Post*, 30 November 2015, https://www.washingtonpost.com/world/national-security/

following-us-indictments-chinese-military-scaled-back-hacks-on-american-industry/2015/11/30/fcdb097a-9450–11e5-b5e4–279b4501e8a6_story.html?utm_term=.41cf15f5946e.

34. AmCham, "American Business in China," 2017 White Paper, https://www.amchamchina.org/about/press-center/amcham-statement/amcham-china-launches-2017-white-paper.

35. Charles Clover, "Foreign Companies in China Hit by New Exchange Controls," *The Financial Times*, 6 December 2016, https://www.ft.com/content/a6d0552a-bbc4–11e6–8b45-b8b81dd5d080.

36. Trend Micro, Inc., "AsiaInfo to Acquire Trend Micro Chinese Subsidiary," press release, 31 August 2015, http://newsroom.trendmicro.com/press-release/financial/asiainfo-acquire-trend-micro-chinese-subsidiary.

37. Jeffrey Knockei et al., "Baidu's and Don'ts: Privacy and Security Issues in Baidu Browser," Monk School of Global Affairs, University of Toronto, 23 February 2016, https://citizenlab.org/2016/02/privacy-security-issues-baidu-browser/.

38. Knockel, Jeffry, *Baidu's and Don't*, Citizen Lab, University of Toronto, 23 February 2006.

39. See Baidu's responses: Citizenlab, "Responses Received from Baidu on February 22nd, 2016," https://citizenlab.org/wp-content/uploads/2016/02/baiduresponses.pdf.

40. David Sanger, "Pentagon Announces New Strategy for Cyberwarfare," *The New York Times*, 23 April 2015.

41. Charlie Mitchell, "McCain Blames Hacks on Admin's Weak Policy," *The Washington Examiner*, 1 February 2016.

Chapter 6

1. Jiaxing and Yangon, "A Tightening Grip," *The Economist*, 14 March 2015, http://www.economist.com/news/briefing/21646180-rising-chinese-wages-will-only-strengthen-asias-hold-manufacturing-tightening-grip.

2. U.S.–China Economic and Security Review Commission, *2015 Annual Report to Congress*, p. 39.

3. Anjani Trevedi, "China, Japan Shed U.S. Treasury Holdings," *The Wall Street Journal*, 18 November 2015, http://www.wsj.com/articles/china-japan-shed-u-s-treasury-holdings-1447824480.

4. Wayne M. Morrison and Marc Labonte, "China's Holdings of U.S. Securities: Implications for the U.S. Economy," Congressional Research

Service, 19 August 2013, https://www.fas.org/sgp/crs/row/RL34314.pdf.

5. Eva Dou, "China's Tech Rules Make It Hard for U.S. Firms to Take Control," *The Wall Street Journal*, 2 June 2016, http://www.wsj.com/articles/chinas-new-tech-rules-make-it-hard-for-u-s-firms-to-take-control-1464870481.

6. Don Clark, "Intel to Convert Processor Chip Factory in China to Make Memory Chips," *The Wall Street Journal*, 20 October 2015, http://www.wsj.com/articles/intel-to-convert-processor-chip-factory-in-china-to-make-memory-chips-1445368532.

7. Pillsbury Winthrop Shaw Pittman, LLP, "China Client Alert Corporate and Securities," accessed 15 April 2016, https://www.pillsburylaw.com/siteFiles/Publications/AlertMar2015Corp_SecuritiesChinasNewForeignInvestmentGuidanceCatalogue.pdf.

8. See Ministry of Commerce, Government of China, "Catalogue for the Guidance of Foreign Investment Industries (Amended in 2011)," http://english.mofcom.gov.cn/article/policyrelease/aaa/201203/20120308027837.shtml, for a complete list of restrictions on investment.

9. Samn Sacks, "Regulatory Barriers to Digital Trade in China, and Costs to US Firms," testimony before the U.S.–China Economic and Security Review Commission, 15 June 2015.

10. Holly Ellyatt, "China Has Conducted a 'War'—not Trade—with Steel, Experts Say," CNBC, 20 May 2016, http://www.cnbc.com/2016/05/20/china-steel-overcapacity-war.html.

11. Josh Horowitz, "Carl Icahn Sold His Apple Stake Because He Is Worried About China's 'Dictatorship' Government," *Quartz*, 29 April 2016, http://qz.com/673035/carl-icahn-sold-his-apple-stake-because-he-is-worried-about-chinas-dictatorship-government/.

12. Jon DiMaggio, "Suckfly: Revealing the Secret Life of Your Code Signing Certificates," Symantec, Inc., 15 March 2016, http://www.symantec.com/connect/blogs/suckfly-revealing-secret-life-your-code-signing-certificates.

13. U.S.–China Economic and Security Review Commission, *2016 Annual Report to Congress*, p. 4.

14. Loren Thompson, "Boeing to Build Its First Offshore Plane Factory in China as Ex-Im Bank Withers," *Forbes*, 15 September 2015.

15. David Barboza, Drew, Christopher and Lohr, Steve, "G.E. to Share Jet Technology with China in New Joint Venture," *The New York Times*, 17 January 2011.

16. John Bussey, "China Venture Is Good for GE but Is It Good for U.S.?" *The Wall Street Journal*, 30 September 2011.

17. Ibid.

18. Andrew Horansky, "Clinton Addresses Cleveland's Lead Problem, Calls Out Eaton Corp," WKYC News, Eaton Corporation, 9 March 2016, http://www.eaton.com/Eaton/OurCompany/NewsEvents/NewsReleases/PCT_264323.

19. Rockwell Collins, Press release, 2012, https://www.rockwellcollins.com/Data/News/2012_Cal_Yr/CS/FY13CSNR01-CETCA-JV.aspx.

20. Business Wire, press release, 12 August 2009, http://www.businesswire.com/news/home/20090812005442/en/Goodrich-China%E2%80%99s-XAIC-Agree-Form-Joint-Venture.

21. Business Wire, press release, 19 February 2016, http://www.businesswire.com/news/home/20160219005941/en/Parker-Aerospace-Chinese-Aerospace-Joint-Ventures-Receive.

22. Stanley Chao, "Viewpoint: Chinese Have Wrong Experience for C919," *Aviation Week and Space Technology*, 6 July 2013.

23. Jim Puzzanghera, "U.S. Prevails in WTO Dispute with China over Auto Tariffs," *The Los Angeles Times*, 23 May 2014.

24. Rose Ru, "Auto Giants Curb Ambitions as China Exits Fast Lane," *The Wall Street Journal*, 25 April 2016, http://www.wsj.com/articles/auto-giants-curb-ambitions-as-china-exits-fast-lane-1461555366.

25. Lawrence Urrich, "Chinese-Made Cars Arrive in U.S. Showrooms," *The New York Times*, 28 January 2016.

26. Gilles Guillaume, "China Car Sales Top U.S." Reuters, 11 January 2010.

27. Dr. Robert D Atkinson, "China's Technological Rise: Challenges to U.S. Innovation and Security," testimony before the House Committee on Foreign Affairs Subcommittee on Asia and the Pacific, 26 April 2017, http://docs.house.gov/meetings/FA/FA05/20170426/105885/HHRG-115-FA05-Wstate-AtkinsonR-20170426.pdf.

28. Michael Rapoport, "SEC Files Suit Against Third Chinese Company," *The Wall Street Journal*, 12 April 2012.

29. James Areddy, Sky Canaves, and Shai Oster, "Rio Tinto Arrests Throw Firms Off Balance," *Wall Street Journal*, 13 August 2009; USA Today, "China Closes 13 Walmart Stores and Arrests 2 Employees," http://www.usatoday.com/money/industries/retail/story/2011-10-13/China-Wal-Mart/50751382/1.

30. Lee Levkowitz, Martella McLellan Ross and J.R. Warner, "The 88 Queensway Group: A Case Study in Chinese Investors' Operations in Angola and Beyond," U.S.-China Economic & Security Review Commission, 10 July 2009/

31. Select Committee of the U.S. House of Representatives, "U.S. National Security and Military/Commercial Concerns with the People's Republic of China," June 2005.

32. Jamil Anderlini, "China's Security Supreme Caught in Bo Fallout," *Financial Times*, 21 April 2012.

33. Richard McGregor and Kathrin Hillel, "Censors Hobbled by Site Outside Great Firewall," *Financial Times*, 23 April 2012.

34. France 24, http://www.france24.com/en/20110811-french-wines-victim-chinese-counterfeiting-chateau-lafite-bordeaux-china-labels.

35. CBS, "The World's Greatest Fakes," 26 January 2004, http://www.cbsnews.com/stories/2004/01/26/60II/main595875.shtml.

36. David Dollar, "China Is Struggling to Keep Its Currency High, Not Low," Brookings Institute, 26 January 2017, https://www.brookings.edu/blog/order-from-chaos/2017/01/26/china-is-struggling-to-keep-its-currency-high-not-low/.

37. Dave Lyons, "China's Golden Shield Project, Myths, Realities and Context," Scribd, http://www.scribd.com/doc/15919071/Dave-Lyons-Chinas-Golden-Shield-Project.

38. David Kravets, "Feds Say China's Net Censorship Imposes Barriers to Free Trade," *Wired*, 20 October 2011, http://www.wired.com/threatlevel/2011/10/china-censorship-trade-barrier/.

39. Loretta Chou and Owen Fletcher, "China Looks at Baidu," *The Wall Street Journal*, 16 September 2011.

40. U.S.-China Economic & Security Review Commission Staff Report, "National Security Implications of Investments and Products from the People's Republic of China in the Telecommunications Sector," staff report, January 2011.

41. Stephanie Kirchgaessner, and Paul Taylor, "Security concerns hold back Huawei," *The Financial Times*, 8 July 2010 https://www.ft.com/content/6fd9f072-8aba-11df-8e17-00144feab49a.

42. Robert Herbold, "In Praise of Chinese Central Planning." Real Clear Markets, 9 July 2011, http://www.realclearmarkets.com/2011/07/09/in_praise_of_chinese_central_planning_115701.html.

43. Robert Herbold, "China vs. America: Which Is the Developing Country?" *The Wall Street Journal*, 9 July 2011.

44. James Mulvenon, "To Get Rich Is Unprofessional: Chinese Military Corruption in the Jiang Era," Hoover Institute, http://media.hoover.org/sites/default/files/documents/clm6_jm.pdf.

45. Worldnet Daily, "America's China Syndrome," 19 August 2003, http://www.wnd.com/?pageId=20358#ixzz1b2qADEoL; see also Jeffrey

Lewis, "How Many Chinese Front Companies?" August 2005, http://lewis.armscontrolwonk.com/archive/727/how-many-chinese-front-companies. Lewis says the original number came from the Cox Report in 1999 and, when questioned about it, Cox said "some of the 3000" were front companies.

46. Joseph Kahn, "Chinese General Threatens Use of A-Bombs If U.S. Intrudes." *The New York Times*, 15 July 2005.

47. Donald Rumsfeld, *Known and Unknown* (New York: Penguin, 2011), p. 316.

Chapter 7

1. Department of Defense, "Annual Report to Congress: Military and Security Developments Involving the People's Republic of China," 2011.

2. Joseph Kahn, "Chinese General Threatens Use of A-Bombs if U.S. Intrudes," *The New York Times*, 15 July 2005.

3. U.S.–China Economic and Security Review Commission, "Capability of the People's Republic of China to Conduct Cyber Warfare and Computer Network Exploitation," 9 October 2009.

4. Steven Levy, "Inside Google's China Misfortune," CNNMoney, 15 April 2011.

5. Keith Johnson, "What Kind of Game Is China Playing?" *Wall Street Journal*, 11 June 2011.

6. Munk Centre for International Studies, Citizen Lab, Shadowserver Foundation, and Information Warfare Monitor, *Shadows in the Cloud: Investigating Cyber Espionage 2.0* ([Toronto, Ont.]: [Citizen Lab, Munk Centre for International Studies, University of Toronto], 2010).

7. Viswanatha, Aruna, *Patriot or Double Agent? CIA Officer on Trial as U.S. Targets Spying by China, The Wall Street Journal*, 22 May 2018, https://www.wsj.com/articles/cia-agent-who-cultivated-ties-with-china-set-for-espionage trial152690400l?emailToken=50af81edce00b5b6 ca7e1983c413098b4DgRsE4qXkewIUOuiC7E65 p39i76zVRsyRbHFBEGv4s0Mu8j%2BDVUp KiiSRGLAM38wip8pPCd%2BlkAizbYZJb56W uhO%2Bg9%2FV0aYB8fJjvvWwrLGlwTrK%2B 6lps%2F%2Bevs2Ea1

8. William C. Hannas, James Mulvenon, and Anna B. Puglisi, *Chinese Industrial Espionage: Technology Acquisition and Military Modernization*, (Routledge, 2013), page 2-3, These authors say the Chinese do not favor ethic Chinese as spies, but follow the same techniques as others spies in their recruiting.

9. Dan Goodin, "Virulent WCry Ransomware Worm May Have North Korea's Fingerprints on It," *Ars Technica*, 15 May 2017, https://arstechnica.com/security/2017/05/virulent-wcry-ransomware-worm-may-have-north-koreas-fingerprints-on-it/.

10. *The Security Ledger*, "Did NSA Hackers The Shadow Brokers Have a Broker?" by "Paul," 19 December 2016, https://securityledger.com/2016/12/did-nsa-hackers-the-shadow-brokers-have-a-broker/.

11. Rachel Chang, "Here's What China's Middle Classes Really Earn—and Spend," Bloomberg L.P., 9 March 2016, https://www.bloomberg.com/news/articles/2016-03-09/here-s-what-china-s-middle-class-really-earn-and-spend.

12. Commission on the Theft of American Intellectual Property, *The IP Commission Report*, National Bureau of Asian Research, May 2013.

13. U.S. Justice Department Indictment, United States of America v. Internet Research Agency LLC, et.al. https://www.justice.gov/file/1035477/

14. Snegovaya, Maria, *Putin's Information Warfare in Ukraine*, The Institute for the Study of War, September 2015, pp. 10–11.

15. Andrew, Christopher and Vasili Mitrokhin, The Sword and the Shield, (Basic Books, New York, New York), 1999, page 243

16. Christopher Andrew, and Gordievsky, KGB, pp. 590–591.

Chapter 8

1. Robert Faris, Hal Roberts, and Stephanie Wang, "China's Green Dam: The Implications of Government Control Encroaching on the Home PC," The OpenNet Initiative, 12 June 2009, https://opennet.net/sites/opennet.net/files/GreenDam_bulletin.pdf.

2. Swati Khandelwal, "Superfish-Like Vulnerability Found in Over 12 More Apps," The Hacker News, 23 February 2015, http://thehackernews.com/2015/02/superfish-vulnerability.html.

3. Software Engineering Institute, Carnegie Mellon University, "Vulnerability Note VU# 870761," 24 November 2015.

4. Adam Langley, "Maintaining Digital Certificate Security," Google Security Blog, 23 March 2015.

5. Jeffrey Knockel, Adam Senft, Ron Deibert, "WUP! There It Is: Privacy and Security Issues in QQ Browser," see also Knockel, Jeffrey, et al., "Baidu's and Don'ts: Privacy and Security Issues in Baidu Browser," 28 March 2016, "Summary: Privacy and Security Issues with UC Browser," 21 May 2015, https://citizenlab.org/?s=Chinese+browse.

6. Chris Smith, "Gartner: Over 1.1 Billion Android Devices to Be Sold in 2014," *The Android Authority*, 7 January 2014, http://www.androidauthority.com/gartner-over-1-1-billion-android-devices-sold-2014-331724/.

7. Dan Farbar, "Google Search Scratches Its Brain 500 Million Times a Day," CNET, 12 May 2013.

8. Sara Radicati and Justin Levenstein, "Email Statistics Report, 2013–2017," The Radicati Group, Inc., April 2013.

9. ArborSert, "ASERT Threat Intelligence Report 2016–03: The Four-Element Sword Engagement," 2016, https://www.arbornetworks.com/blog/asert/wp-content/uploads/2016/04/ASERT-Threat-Intelligence-Report-2016-03-The-Four-Element-Sword-Engagement.pdf.

10. Andrei Soldatov and Irina Borogan, *The Red Web* (New York: Public Affairs, 2015), pp. 185–190.

11. Leonid Ragozin and Michael Riley, "Putin Is Building a Great Russian Firewall," Bloomberg Businessweek, 25 August 2016, https://www.bloomberg.com/news/articles/2016-08-26/putin-is-building-a-great-russian-firewall.

12. Max Seddon, "Russia's Chief Internet Censor Enlists China's Know-How," *The Financial Times*, 26 April 2016, https://www.ft.com/content/08564d74-0bbf-11e6-9456-444ab52lla2f.

13. Yukyung Yeo, "Regulatory Politics in China's Telecommunications Service Industry: When Socialist Market Economy Meets Independent Regulator Model," June 2008, http://regulation.upf.edu/utrecht-08-papers/yyeo.pdf.

14. Jose Pagliery, "Ex-NSA Director: China Has Hacked 'Every Major Corporation' in U.S.," CNN Money, 16 March 2015, http://money.cnn.com/2015/03/13/technology/security/chinese-hack-us/.

15. George Chen, FP Feb 2015.

16. Greatfire.org, "GitHub Blocked in China—How It Happened, How to Get Around It, and Where It Will Take Us," by "Percy," 23 January 2013, https://en.greatfire.org/blog/2013/jan/github-blocked-china-how-it-happened-how-get-around-it-and-where-it-will-take-us.

17. Bill Marczak et al., "China's Great Cannon," University of Toronto, 10 April 2015, https://citizenlab.org/2015/04/chinas-great-cannon/.

18. Greg Walton, "China's Golden Shield: Corporations and the Development of Surveillance Technology in the People's Republic of China," International Centre for Human Rights and Democratic Development, 2001.

19. David E. Sanger, "U.S. Decides to Retaliate Against China's Hacking," *The New York Times*, 31 July 2015, http://www.nytimes.com/2015/08/01/world/asia/us-decides-to-retaliate-against-chinas-hacking.html?_r=0.

20. Zunyou Zhou, "China's Comprehensive Counter-Terrorism Law," *The Diplomat*, 23 January 2016, http://thediplomat.com/2016/01/chinas-comprehensive-counter-terrorism-law/.

21. Paul Mozur and Jane Perlez, "China Bets on Sensitive U.S. Start-Ups, Worrying the Pentagon," *The New York Times*, 22 March 2017, http://mobile.nytimes.com/2016/05/17/technology/china-quietly-targets-us-tech-companies-in-security-reviews.html.

22. Paul Mozur and Jack Ewing, "Rush of Chinese Investment in Europe's High-Tech Firms Is Raising Eyebrows," *The New York Times*, 16 September 2016, http://www.nytimes.com/2016/09/17/business/dealbook/china-germany-takeover-merger-technology.html?_r=1.

23. Aixtron.com, "Aixtron Transaction Fact Sheet: Aixtron SE," May 2016, http://www.aixtron.com/fileadmin/user_upload/IR/2016/Fact_Sheet_-_Transaction_Fact_Sheet_EN.pdf.

24. The Infosec Institute, "Panama Papers: How Hackers Breached the Mossack Fonseca Firm," posted 20 April 2016, http://resources.infosecinstitute.com/panama-papers-how-hackers-breached-the-mossack-fonseca-firm/.

25. BBC News, "Panama Papers: China Leaders' Relatives Named in Leaks," 4 April 2016, http://www.bbc.com/news/world-asia-35962326.

26. Neil Chenoweth, "The Panama Papers: Chinese Rich Listers Were Top Australian Clients," *The Australian Financial Review*, 9 May 2016, http://www.afr.com/news/policy/tax/the-panama-papers-chinese-rich-listers-were-top-australian-clients-20160508-gop5b1.

27. Marina Koren, "Panama Papers: The Mossack Fonseca Investigations Begin," *The Atlantic*, 13 April 2016, http://www.theatlantic.com/international/archive/2016/04/panama-papers-mossack-fonseca-raid/478014/.

28. The Asian Age, "China's ZTE Executives to Step Down," 4 April 2016, http://dailyasianage.com/news/15162/chinas-zte-executives-to-step-down.

29. World Trade Organization, "International Trade Statistics 2015," p. 25.

30. Christopher S. Wren, "Insurers Swindled Jews, Nazi Files Show," *The New York Times*, 18 May 1998.

31. Krebs on Security, "Who Else Was Hit by the RSA Attackers," http://krebsonsecurity.com/2011/10/who-else-was-hit-by-the-rsa-attackers/.

32. Johathan Weisman, "U.S. to Share Cautionary Tale of Trade Secret Theft with Chinese Official," *The New York Times*, 15 February 2010, p. A-10.

33. Carl Meyer, "Are Chinese Spies Getting an Easy Ride?" *Embassy*, July 2011, http://www.embassymag.ca/page/printpage/spies-07-27-2011.

34. Cory Bennett, "NSA Head: China Still Spying on U.S. Companies," *The Hill*, 5 April 2016.

35. Robert Hackett, "China's Cyber Spying on the U.S. Has Drastically Changed," Fortune, Inc., 25 June 2016.

36. Shirley Kan, "China: Suspected Acquisition of U.S. Nuclear Weapon Secrets," Congressional Research Services, February 2006.

37. Select Committee of the U.S. House of Representatives, "U.S. National Security and Military/Commercial Concerns with the People's Republic of China," June 2005.

38. Ibid.

39. Stewart D. Personick and Cynthia A. Patterson, eds., National Research Council of the National Academies, *Critical Information Infrastructure Protection and the Law* (Washington, D.C.: National Academy of Engineering, National Academies Press, 2003), p. 15.

40. SiobHan Gorman, "Electricity Grid in U.S. Penetrated by Spies," *Wall Street Journal*, 8 April 2009.

41. U.S.–China Economic and Security Review Commission, *2010 Annual Report to Congress*, 111th Congress, 2nd Session, November 2010.

42. Carl Von Clausewitz, *On War* ([S.L.]: Value Classic Reprints, 2017) p. 102.

43. http://www.networkworld.com/news/2006/102306counterfeit.html. This page has since been removed.

44. Phyllis Schlafly, "Buying Counterfeit Chips from China," TownHall, 4 October 2011, http://townhall.com/columnists/phyllisschlafly/2011/10/04/buying_counterfeit_chips_from_china/page/full/.

45. U.S.–China Economic and Security Review Commission, "National Security Implications of Investments and Products from the People's Republic of China in the Telecommunications Sector," staff report, January 2011.

46. Aleksandr Solzhenitsyn, *The Gulag Archipelago* (New York: Collins, 1974), p. 168.

47. Clay Wilson, "High Altitude Electromagnetic Pulse (HEMP) and High-Power Microwave Devices: Threat Assessments," 21 July 2008, http://www.fas.org/sgp/crs/natsec/RL32544.pdf.

Chapter 9

1. Kathrin Hille, "A Show of Force: China's Military," *Financial Times*, 30 September 2011.

2. Fox News, "French Wines Victim of Chinese Counterfeiting," http://video.foxnews.com/v/5269758235001/?#sp=watch-live.

3. Mary Cawte, *Making Radio into a Tool for War*, research paper for the Australian Research Council, 1996, [SOURCE OR REPOSITORY], p. 8.

4. Anonymous, *Agent Radio Operation during World War II*, Central Intelligence Agency Library, approved for public release September 1993, https://www.cia.gov/library/center-for-the-study-of-intelligence/kent-csi/vol3no1/html/v03i1a10p_0001.htm.

5. James Eng, "Iran-Based Hackers Created Network of Fake Linkedin Profiles," NBC News, 7 October 2015, http://www.cnbc.com/2015/10/07/iran-based-hackers-created-network-of-fake-linkedin-profiles-report.html.

6. Reuters, "Facebook Says It Will Battle Disinformation Operations," *Fortune*, 27 April 2017, http://fortune.com/2017/04/27/facebook-disinformation-operations/.

7. Chad P. Brown and Alan O. Sykes, "The Trump Trade Team's Vocabulary Problem," *The Wall Street Journal*, 14 May 2017, https://www.wsj.com/article_email/the-trump-trade-teams-vocabulary-problem-1494795625-lMyQjAxMTA3NjElNTgxOTU2Wj/.

8. Murong Xecun, "Scaling China's Great Firewall," *The New York Times*, 17 August 2015.

9. Gary King, Jennifer Pan, and Margaret E. Roberts, "Reverse-Engineering Censorship in China: Randomized Experimentation and Participant Observation," *Science Magazine* 345, no. 6199 (22 August 2014).

10. Ibid.

11. James R. Clapper, "Worldwide Threat Assessment of the US Intelligence Community," Statement for the Record. Senate Select Committee on Intelligence. 9 February 2016, p. 3.

12. *Washington Post*, "Will China Keep Its Cyber Promises?" 21 October 2015.

Bibliography

Agent Radio Operation During World War II. Central Intelligence Agency Library. Approved for public release September 1993. https://www.cia.gov/library/center-for-the-study-of-intelligence/kent-csi/vol3no1/html/v03i1a10p_0001.htm.

Aghekyan, Elen, Bret Nelson, et al. *Freedom of the Press 2016.* Washington, D.C.: Freedom House, 2016.

AmCham. "American Business in China." 2017 White Paper. https://www.amchamchina.org/about/press-center/amcham-statement/amcham-china-launches-2017-white-paper.

Anderlini, Jamil. "China's Security Supreme Caught in Bo Fallout." *Financial Times,* 21 April 2012.

Areddy, James, Sky Canaves, and Shai Oster. "Rio Tinto Arrests Throw Firms Off Balance." *The Wall Street Journal,* 13 August 2009.

Arms Control Association. "U.S. Missile Defense Programs at a Glance." August 2016. https://www.armscontrol.org/factsheets/usmissiledefense.

Asawa, Juro. "ZTE to Replace Three Senior Executives." *The Wall Street Journal,* 2 April 2016.

Asia Maritime Transparency Initiative. "Airpower Projection." http://amti.csis.org/

Associated Press. "Lithuanian 'Elves' Combat Russian Influence On Line." 28 December 2016. http://www.foxbusiness.com/features/2016/12/28/lithuanian-elves-combat-russian-influence-online.html.

Atkinson, Robert D. "China's Technological Rise: Challenges to U.S. Innovation and Security." Testimony before the House Committee on Foreign Affairs Subcommittee on Asia and the Pacific, 26 April 2017. http://docs.house.gov/meetings/FA/FA05/20170426/105885/HHRG-115-FA05-Wstate-AtkinsonR-20170426.pdf.

Baker, Steward. "NSA Files." *The Guardian.* Accessed November 2015. http://www.theguardian.com/world/interactive/2013/nov/01/snowden-nsa-files-surveillance-revelations-decoded#section/7.

Baocun, Wang (senior colonel) and Li Fei. "Information Warfare," summarized from articles in the *Liberation Army Daily,* 1995.

Barboza, David, Christopher Drew, and Steve Lohr. "G.E. to Share Jet Technology with China in New Joint Venture." *The New York Times,* 17 January 2011.

BBC. "China's First Aircraft Carrier Starts First Sea Trials." 10 August 2011. http://www.bbc.co.uk/news/world-asia-pacific-14470882.

_____. "Kung Fu Grandma Is China's New Internet Sensation." http://www.bbc.com/news/video_and_audio/headlines/39073334.

_____. "MH17 Ukraine Plane Crash: What We Know." 28 September 2016. http://www.bbc.com/news/world-europe-28357880.

_____. "Migrant Crisis: Migration to Europe Explained in Seven Charts." 4 March 2016. http://www.bbc.com/news/world-europe-34131911.

Bennett, Cory. "Guccifer 2.0 Drops More DNC Docs." *Politico,* 13 September 2016.

Benyumov, Konstantin. "How Russia's Independent Media Was Dismantled Piece by Piece." *The Guardian,* 25 May 2016. https://www.theguardian.com/world/2016/may/25/how-russia-independent-media-was-dismantled-piece-by-piece.

Black, Henry Campbell. *Black's Law Dictionary.* 6th ed. St. Paul, MN: West, 1990.

Blanchard, Ben, and Benjamin Kang Lim. "'Give Them a Bloody Nose': Xi Pressed for Stronger South China Sea Response." Reuters, 31 July 2016. http://www.reuters.com/article/us-south-chinasea-ruling-china-insight-idUSKCN10B10G.

Boot, Max, Jeane J. Kirkpatrick, et al. "Political Warfare." Council on Foreign Relations, June 2013. http://www.cfr.org/wars-and-warfare/political-warfare/p30894.

British Broadcasting Service. "Democrat Hack: Who Is Guccifer 2.0?" 28 July 2016.

Brown, Chad P., and Alan O. Sykes. "The Trump Trade Team's Vocabulary Problem." *The Wall Street Journal*, 14 May 2017. https://www.wsj.com/article_email/the-trump-trade-teams-vocabulary-problem-1494795625-lMyQjAxMTA3NjE1NTgxOTU2Wj/.

Browne, Andrew. "Man in the Middle: Rodrigo Duterte Gets a Taste of China's Heavy Hand." *The Wall Street Journal*, 19 July 2016. http://www.wsj.com/articles/man-in-the-middle-rodrigo-duterte-gets-a-taste-of-china-1468909553.

Burke, Garance, and Jonathan Fahey. "Iranian Hackers Breached U.S. Power Grid to Engineer Blackouts." *The Times of Israel*, 22 December 2015. http://www.timesofisrael.com/iranian-hackers-breached-us-power-grid-to-engineer-blackouts/.

Business Wire. Press release, 12 August 2009. http://www.businesswire.com/news/home/20090812005442/en/Goodrich-China%E2%80%99s-XAIC-Agree-Form-Joint-Venture.

_____. Press release, 19 February 2016. http://www.businesswire.com/news/home/20160219005941/en/Parker-Aerospace-Chinese-Aerospace-Joint-Ventures-Receive.

Bussey, John. "China Venture Is Good for GE but Is It Good for U.S.?" *The Wall Street Journal*, 30 September 2011.

Caryl, Christian. "Novorossiya Is Back from the Dead." *Foreign Policy*, 17 April 2014. http://foreignpolicy.com/2014/04/17/novorossiya-is-back-from-the-dead/.

Cawte, Mary. "Making Radio into a Tool for War." Research paper for the Australian Research Council, 1996.

CBS. "The World's Greatest Fakes." 26 January 2004. http://www.cbsnews.com/stories/2004/01/26/60II/main595875.shtml.

CCTV Business Journal. Video. 7 September 2012. https://www.youtube.com/watch?v=OiZm2hIMDs4&feature=youtu.be.

CCTV Global Business. "China's Dandong Port Co. Could Benefit from Democratic National Convention." 7 September 2012. https://www.youtube.com/watch?v=OiZm2hIMDs4.

Chao, Stanley. "Viewpoint: Chinese Have Wrong Experience for C919." *Aviation Week and Space Technology*, 6 July 2013.

Chin, Josh. "Cyber Sleuths Track Hacker to China's Military." *The Wall Street Journal*, 23 September 2015.

Choe, San-Hun. "Computer Networks in South Korea Are Paralyzed in Cyberattacks." *The New York Times*, 20 March 2013.

Chou, Loretta, and Owen Fletcher. "China Looks at Baidu." *The Wall Street Journal*, 16 September 2011.

Churchill, Winston. "The Churchill Society, Churchill's Wartime Speeches, Excerpt from 'The Munich Agreement: A Total and Unmitigated Defeat.'" House of Commons, 5 October 1938. http://www.churchill-society-london.org.uk/Munich.html.

CIA. "Comments on [redacted] Study 'The Vulnerability of the Soviet Union and Its European Satellites to Political Warfare.'" Central Intelligence Agency Redacted Report, declassified 5 June 2013. https://www.cia.gov/library/readingroom/docs/CIA-RDP61S00750A000300100055-6.pdf.

Citizenlab. "Responses Received from Baidu on February 22nd, 2016." https://citizenlab.org/wp-content/uploads/2016/02/baiduresponses.pdf.

Clapper, James R. "Worldwide Cyber Threats." Statement for the Record. House Permanent Select Committee Intelligence, 10 September 2015.

_____. "Worldwide Threat Assessment of the US Intelligence Community." Statement for the Record. Senate Select Committee on Intelligence, 9 February 2016.

Clark, Don. "Intel to Convert Processor Chip Factory in China to Make Memory Chips." *The Wall Street Journal*, 20 October 2015. http://www.wsj.com/articles/intel-to-convert-processor-chip-factory-in-china-to-make-memory-chips-1445368532.

Clausewitz, Carl Von. *On War*. [S.L.]: Value Classic Reprints, 2017.

Clover, Charles. "Foreign Companies in China Hit by New Exchange Controls." *The Financial Times*, 6 December 2016. https://www.ft.com/content/a6d0552a-bbc4-11e6-8b45-b8b81dd5d080.

Colucci, Nicholas. "Testimony Before the Senate Judiciary Committee." 2 February 2016. https://www.judiciary.senate.gov/imo/media/doc/02-02-16%20Colucci%20Testimony.pdf.

Confessore, Nicholas, and Stephanie Saul. "Inquiry Highlights Terry McAuliffe's Ties to Chinese Company." *The New York Times*, 24 May 2016. https://www.nytimes.com/2016/05/25/us/politics/terry-mcauliffe-wang-wenliang.html.

Deibert, Ron. "Cyber Crime and Warfare." University of Toronto. Accessed November 2015. https://www.youtube.com/embed/PqvKGfDXFE4?list=UUu_u-P3cBFO7D-sAjxd_I-w.

Department of Commerce, Bureau of Industry and Security. "Proposal for Import and Export

Control Risk Avoidance." Internal document of ZTE posted in English at https://www.bis.doc.gov/index.php/forms-documents/about-bis/newsroom/1436-proposal-for-english/file.

Department of Defense. "Joint Publication 3–13, Information Operations." 20 November 2014.

DiMaggio, Jon. "Suckfly: Revealing the Secret Life of Your Code Signing Certificates." Symantec, Inc., 15 March 2016. http://www.symantec.com/connect/blogs/suckfly-revealing-secret-life-your-code-signing-certificates.

Dizon, Nikko, and Nina P. Callega. "PH: China '9-Dash Line' Doesn't Exist." *Philippine Daily Inquirer*, 24 November 2015.

Dollar, David. "China Is Struggling to Keep Its Currency High, Not Low." Brookings Institute, 26 January 2017. https://www.brookings.edu/blog/order-from-chaos/2017/01/26/china-is-struggling-to-keep-its-currency-high-not-low/.

Dou, Eva. "China to Start Security Checks on Technology Companies in June." *The Wall Street Journal*, 3 May 2017. https://www.wsj.com/articles/china-to-start-security-checks-on-technology-companies-in-june-1493799352.

_____. "China's Tech Rules Make It Hard for U.S. Firms to Take Control." *The Wall Street Journal*, 2 June 2016. http://www.wsj.com/articles/chinas-new-tech-rules-make-it-hard-for-u-s-firms-to-take-control-1464870481.

Ellyatt, Holly. "China Has Conducted a 'War'—not Trade—with Steel, Experts Say." CNBC, 20 May 2016. http://www.cnbc.com/2016/05/20/china-steel-overcapacity-war.html.

Eng, James. "Iran-Based Hackers Created Network of Fake Linkedin Profiles." NBC News, 7 October 2015. http://www.cnbc.com/2015/10/07/iran-based-hackers-created-network-of-fake-linkedin-profiles-report.html.

Feast, Lincoln, and Greg Torode. "Exclusive: Risking Beijing's ire, Vietnam Begins Dredging on South China Sea Reef." Reuters, 9 December 2016. http://www.reuters.com/article/us-southchinasea-vietnam-idUSKBN13X0WD.

Federation of American Scientists. "2018 Nuclear Posture Review Resource." http://fas.org/nuke/guide/dprk/missile/td-2.htm.

FireEye Threat Intelligence. "Hammertoss: Stealthy Tactics Define a Russian Cyber Threat Group." 29 July 2015. https://www.fireeye.com/blog/threatresearch/2015/07/hammertoss_stealthy.html.

Fox News. "French Wines Victim of Chinese Counterfeiting." http://video.foxnews.com/v/5269758235001/?#sp=watch-live.

France 24. http://www.france24.com/en/20110811-french-wines-victim-chinese-counterfeiting-chateau-lafite-bordeaux-china-labels.

Francis, David. "U.S. Treasury Hits Russia with More Sanctions over Ukraine." *Foreign Policy*, 1 September 2016. http://foreignpolicy.com/2016/09/01/u-s-treasury-hits-russia-with-more-sanctions-over-ukraine/.

Groll, Elias. "'Obama's General' Pleads Guilty to Leaking Stuxnet Operation." *Foreign Policy*, 17 October 2016. http://foreignpolicy.com/2016/10/17/obamas-general-pleads-guilty-to-leaking-stuxnet-operation/.

The Guardian. "Agence France-Presse in Berlin, Russia Accused of Series of International Cyber-Attacks." 13 May 2016.

The Guardian. "Agence France-Presse, South China Sea: Beijing Tells G7 Foreign Ministers to Keep Out of Territorial Dispute." 12 April 2016.

Guillaume, Gilles. "China Car Sales Top U.S." Reuters, 11 January 2010.

Herbold, Robert. "China vs. America: Which Is the Developing Country?" *The Wall Street Journal*, 9 July 2011.

_____. "In Praise of Chinese Central Planning." Real Clear Markets, 9 July 2011. http://www.realclearmarkets.com/2011/07/09/in_praise_of_chinese_central_planning_115701.html.

Hille, Kathrin. "A Show of Force: China's Military." *Financial Times*, 30 September 2011.

Horansky, Andrew. "Clinton Addresses Cleveland's Lead Problem, Calls Out Eaton Corp." WKYC News, Eaton Corporation, 9 March 2016. http://www.eaton.com/Eaton/OurCompany/NewsEvents/NewsReleases/PCT_264323.

Horowitz, Josh. "Carl Icahn Sold His Apple Stake Because He Is Worried About China's 'Dictatorship' Government." *Quartz*, 29 April 2016. http://qz.com/673035/carl-icahn-sold-his-apple-stake-because-he-is-worried-about-chinas-dictatorship-government/.

Hsu, Stacy. "Presidential Office Rejects Criticism." *Taipei Times*, 11 May 2016. http://www.taipeitimes.com/News/taiwan/archives/2016/05/11/2003646004.

Ilves, Toomas Hendrik. "Prepared Testimony: Undermining Democratic Institutions and Splintering NATO: Russian Disinformation." The House Foreign Affairs Committee, March 9, 2017. http://docs.house.gov/meetings/FA/FA00/20170309/105674/HHRG-115-FA00-Wstate-IlvesH-20170309.pdf.

Jiaxing and Yangon. "A Tightening Grip." *The Economist*, 14 March 2015. http://www.economist.com/news/briefing/21646180-rising-chinese-wages-will-only-strengthen-asias-hold-manufacturing-tightening-grip.

Johnson, Jesse. "China Deploys Anti-Diver Rocket Launchers to Man-Made Island in South China Sea: Report." *The Japan Times*, undated article. http://www.japantimes.co.jp/news/2017/05/17/asia-pacific/china-deploys-anti-diver-rocket-launchers-man-made-island-south-china-sea-report/#.WR8i4j OZOL8.

Kahn, Joseph. "Chinese General Threatens Use of A-Bombs if U.S. Intrudes." *The New York Times*, 15 July 2005.

Kaplan, Fred. *Dark Territory*. New York: Simon and Shuster, 2016.

King, Gary, Jennifer Pan, and Margaret E. Roberts. "Reverse-Engineering Censorship in China: Randomized Experimentation and Participant Observation." *Science Magazine* 345, no. 6199 (22 August 2014).

Klinger, Bruce. "Chinese Foot-Dragging on North Korea Thwarts U.S. Security Interests." The Heritage Foundation, 11 April 2016. http://www.heritage.org/research/reports/2016/04/chinese-foot-dragging-on-north-korea-thwarts-us-security-interests.

Knockel, Jeffrey, et al. "Baidu's and Don'ts: Privacy and Security Issues in Baidu Browser." Monk School of Global Affairs, University of Toronto, 23 February 2016. https://citizenlab.org/2016/02/privacy-security-issues-baidu-browser/.

Kovacs, Eduard. "APT3 Hackers Linked to Chinese Ministry of State Security." *Securityweek*, 17 May 2017. http://www.securityweek.com/apt3-hackers-linked-chinese-ministry-state-security.

Kravets, David. "Feds Say China's Net Censorship Imposes Barriers to Free Trade." *Wired*, 20 October 2011. http://www.wired.com/threatlevel/2011/10/china-censorship-trade-barrier/.

Krebs, Brian. "Who Else Was Hit by the RSA Attackers?" Krebs on Security. http://krebsonsecurity.com/2011/10/who-else-was-hit-by-the-rsa-attackers/.

LaGrone, Sam. "PACOM Harris: U.S. Would Ignore a 'Destabilizing' Chinese South China Sea Air Defense Identification Zone." *U.S. Naval Institute News*, 26 February 2016.

Lake, Eli. "China Bid Blocked over Spy Worry." *Daily Beast*, 11 October 2011.

Lasker, John. "Watchdogs Sniff Out Terror Sites." *Wired News*, 25 February 2005. http://www.wired.com/news/privacy/0,67223-0.html?tw=wn_story_page_prev2].

_____. "U.S. Military's Elite Hacker Crew." *Wired News*, April 18, 2005. http://www.

wired.com/news/privacy/0,1848,67223,00.html.

Lee, Carol E., and Jay Solomon. "U.S. Targets North Korea in Retaliation for Sony Hack." *The Wall Street Journal*, 3 January 2015.

Lewis, Jeffrey. "How Many Chinese Front Companies?" August 2005. http://lewis.armscontrolwonk.com/archive/727/how-many-chinese-front-companies. Lewis says the original number came from the Cox Report in 1999 and, when questioned about it, Cox said "some of the 3000" were front companies.

Lei, Zhao. "New Satellite Keeps Eye on Sea Interests." *China Daily*, 11 August 2016. http://www.chinadaily.com.cn/china/2016-08/11/content_26426328.htm.

Levkowitz, Lee, Martella McLellan Ross, and J.R. Warner. "The 88 Queensway Group: A Case Study in Chinese Investors' Operations in Angola and Beyond." U.S.–China Economic & Security Review Commission, 10 July 2009.

Lewis, Brian C. "Information Warfare." Original source not identified. https://fas.org/irp/eprint/snyder/infowarfare.htm.

Lyons, Dave. "China's Golden Shield Project, Myths, Realities and Context." Scribd. http://www.scribd.com/doc/15919071/Dave-Lyons-Chinas-Golden-Shield-Project.

Martin, Jonathan, and Alan Rappeport. "Debbie Wasserman Schultz to Resign D.N.C. Post." *The New York Times*, 24 July 2016. https://www.nytimes.com/2016/07/25/us/politics/debbie-wasserman-schultz-dnc-wikileaks-emails.html?_r=0.

McDonald-Gibson, Charlotte. "Europe Mulls a Russian Language TV Channel to Counter Moscow Propaganda." *Time*, 19 January 2015. http://time.com/3673548/europe-russian-language-tv/.

McGregor, Richard, and Kathrin Hillel. "Censors Hobbled by Site Outside Great Firewall." *Financial Times*, 23 April 2012.

Melander, Erik. *Encyclopedia of Political Thought*. n.p.: John Wiley, 2014. http://uu.diva-portal.org/smash/record.jsf?pid=diva2%3A799013&dswid=-335.

Ministry of Commerce, Government of China. http://english.mofcom.gov.cn/article/policyrelease/aaa/201203/20120308027837.shtml. A complete list of restrictions on investment.

Mitchell, Amy, et al. "Trust and Accuracy from the Modern News Consumer." Pew Research Center, 7 July 2016. http://www.journalism.org/2016/07/07/trust-and-accuracy/.

Mitchell, Charlie. "McCain Blames Hacks on Admin's Weak Policy." *The Washington Examiner*, 1 February 2016.

Molander, Roger C., Andrew S. Riddile, and Peter A. Wilson. *Strategic Information Warfare*. Santa Monica, CA: Rand, 1996.

Morgenstern, George. "The Actual Road to Pearl Harbor." In *Perpetual War for Perpetual Peace,* ed. Harry Elmer Barnes, pp. 332–343. Caldwell, Idaho: Caxton, 1953.

Morrison, Wayne M., and Marc Labonte. "China's Holdings of U.S. Securities: Implications for the U.S. Economy." Congressional Research Service, 19 August 2013. https://www.fas.org/sgp/crs/row/RL34314.pdf.

Mozur, Paul, and Janie Perlez. "China Bets on Sensitive U.S. Start-Ups, Worrying the Pentagon." *The New York Times*, 22 March 2017. https://www.nytimes.com/2017/03/22/technology/china-defense-start-ups.html.

Mulvenon, James. "To Get Rich Is Unprofessional: Chinese Military Corruption in the Jiang Era." Hoover Institute. http://media.hoover.org/sites/default/files/documents/clm6_jm.pdf.

Munk Centre for International Studies, Citizen Lab, Shadowserver Foundation, and Information Warfare Monitor. *Shadows in the Cloud: Investigating Cyber Espionage 2.0,* [Toronto, Ont.]: [Citizen Lab, Munk Centre for International Studies, University of Toronto], 2010.

Nakashima, Ellen. "Following U.S. Indictments, China Shifts Commercial Hacking away from Military to Civilian Agency." *The Washington Post*, 30 November 2015. https://www.washingtonpost.com/world/national-security/following-us-indictments-chinese-military-scaled-back-hacks-on-american-industry/2015/11/30/fcdb097a-9450-11e5-b5e4-279b4501e8a6_story.html?utm_term=.41cf15f5946e.

_____. "U.S. Developing Sanctions Against China over Cyberthefts." *The Washington Post,* 30 August 2015. https://www.washingtonpost.com/world/national-security/administration-developing-sanctions-against-china-over-cyberespionage/2015/08/30/9b2910aa-480b-11e5-8ab4-c73967a143d3_story.html?utm_term=.d020b3917047.

_____. "When Is a Cyberattack an Act of War?" *The Washington Post,* 26 October 2012. https://www.washingtonpost.com/opinions/when-is-a-cyberattack-an-act-of-war/2012/10/26/02226232-1eb8-11e2-9746-908f727990d8_story.html.

Nolan, Joseph R., et al. *Black's Law Dictionary*. St. Paul, MN: West, 1990.

O'Conner, Sean. "Imagery Shows Chinese HQ-9 Battery Being Removed from Woodly Island." Jane's 360, 21 July 2016. http://www.janes.com/article/62442/imagery-shows-chinese-hq-9-battery-being-removed-from-woody-island.

O'Rourke, Ronald. "Maritime Territorial and Exclusive Economic Zone (EEZ) Disputes Involving China: Issues for Congress." Congressional Research Service, United States Congress, 22 December 2015.

Osawa, Juro, and Eva Dou. "U.S. to Place Trade Restrictions on China's ZTE." *The Wall Street Journal,* 7 March 2016.

Parameswaran, Prashanth. "China Enforcing Quasi-ADIZ in South China Sea: Philippine Justice." *The Diplomat,* 13 October 2015.

Perez, Evan. "U.S. Official Blames Russia for Power Grid Attack in Ukraine." Cable News Network, 11 February 2016. http://www.cnn.com/2016/02/11/politics/ukraine-power-grid-attack-russia-us/.

Pillsbury Winthrop Shaw Pittman, LLP. "China Client Alert Corporate and Securities." Accessed 15 April 2016. https://www.pillsburylaw.com/siteFiles/Publications/AlertMar2015Corp_SecuritiesChinasNewForeignInvestmentGuidanceCatalogue.pdf.

Pompeo, Mike. "Remarks as Prepared for Delivery by Central Intelligence Agency Director Mike Pompeo at the Center for Strategic and International Studies." 13 April 2017. https://www.cia.gov/news-information/speeches-testimony/2017-speeches-testimony/pompeo-delivers-remarks-at-csis.html.

Puzzanghera, Jim. "U.S. Prevails in WTO Dispute with China over Auto Tariffs." *The Los Angeles Times,* 23 May 2014.

Rapoport, Michael. "SEC Files Suit Against Third Chinese Company." *The Wall Street Journal,* 12 April 2012.

Reuters. "Facebook Says It Will Battle Disinformation Operations." *Fortune,* 27 April 2017. http://fortune.com/2017/04/27/facebook-disinformation-operations/.

Rockwell Collins. Press release, 2012. https://www.rockwellcollins.com/Data/News/2012_Cal_Yr/CS/FY13CSNR01-CETCA-JV.aspx.

Roth, Andrew. "Russia and China Sign Cooperation Pacts." *The New York Times,* 8 May 2015. https://www.nytimes.com/2015/05/09/world/europe/russia-and-china-sign-cooperation-pacts.html.

Roth, John. "Investigation into Employee Complaints About the Management of U.S. Citizenship and Immigration Services' EB-5 Program." Inspector General Homeland Security, 24 March 2015. https://www.oig.dhs.gov/assets/Mga/OIG_mga-032415.pdf.

Ru, Rose. "Auto Giants Curb Ambitions as China Exits Fast Lane." *The Wall Street Jour-*

nal, 25 April 2016. http://www.wsj.com/art icles/auto-giants-curb-ambitions-as-china-exits-fast-lane-1461555366.

Rumsfield, Donald. *Known and Unknown*. New York: Penguin, 2011.

Russia RT. "Crimean Energy Bridge Completed from Mainland Russia." 12 May 2016. https://www.rt.com/business/342737-crimea-energy-bridge-complete/.

_____. "Emails Expose Watchdog's Dollar Deals." 8 December 2011. https://www.rt.com/news/election-america-golos-support-393/.

Sacks, Samn. "Regulatory Barriers to Digital Trade in China, and Costs to US Firms." Testimony before the U.S.–China Economic and Security Review Commission, 15 June 2015.

Sainato, Michael. "Exclusive: Wikileaks Guccifer 2.0 Teaser Exposes Pay-to-Play and Financial Data." *The New York Observer*, 5 October 2016. http://observer.com/2016/10/exclusive-wikileaks-guccifer-2-0-teaser-exposes-pay-to-play-and-financial-data/.

Sanger, David. "Pentagon Announces New Strategy for Cyberwarfare." *The New York Times*, 23 April 2015.

Sanger, David E., and Martin Fackler. "N.S.A. Breached North Korean Networks Before Sony Attack, Officials Say." *The New York Times*, 18 January 2015.

Sanger, David E., and Nicole Perlroth. "As Democrats Gather, a Russian Subplot Raises Intrigue." *The New York Times*, 24 July 2016. https://www.nytimes.com/2016/07/25/us/politics/donald-trump-russia-emails.html.

_____. "N.S.A. Breached Chinese Servers Seen as Security Threat." *The New York Times*, 22 March 2014. https://www.nytimes.com/2014/03/23/world/asia/nsa-breached-chinese-servers-seen-as-spy-peril.html.

Scannell, Kara. "FBI Details North Korean Attack on Sony." *The Financial Times*, 7 January 2015. https://www.ft.com/content/287beee4-96a2-11e4-a83c-00144feabdc0.

Schreck, Carl. "Russian Lawyer Says FSB Officers, Kaspersky Manager Charged with Treason." Radio Free Europe/Radio Liberty, 1 February 2017. http://www.rferl.org/a/russia-fsb-officers-treason-kaspersky/28272937.html.

Select Committee of the U.S. House of Representatives. "U.S. National Security and Military/Commercial Concerns with the People's Republic of China." June 2005.

Semuels, Alana. "Should Congress Let Wealthy Foreigners Buy Green Cards?" *The Atlantic*, 21 September 2015. https://www.theatlantic.com/business/archive/2015/09/should-congress-let-wealthy-foreigners-buy-citizenship/406432/.

Sharkov, Damien. "Half of Russians Do Not Trust Russian Media: Poll." *Newsweek*, 17 October 2016. http://www.newsweek.com/majority-russians-do-not-trust-national-media-510682.

Smith, Paul A. *On Political War*. Washington, D.C.: National Defense University Press, 1989.

Strohm, Chris. "North Korea Web Outage Response to Sony Hack, Lawmaker Says." Bloomberg L.P., 17 March 2015. http://www.bloomberg.com/politics/articles/2015-03-17/north-korea-web-outage-was-response-to-sony-hack-lawmaker-says.

Swift, Art. "Americans' Trust in Mass Media Sinks to New Low." Gallup Inc., 14 September 2016. http://www.gallup.com/poll/195542/americans-trust-mass-media-sinks-new-low.aspx.

Tenet, George J. "DCI Testimony Before the Senate Select Committee on Government Affairs." 24 June 1998. https://www.cia.gov/news-information/speeches-testimony/1998/dci_testimony_062498.html.

Thielman, Sam, and Spencer Ackerman. "Cozy Bear and Fancy Bear: Did Russians Hack Democratic Party and If So, Why?" *The Guardian*, 16 July 2016. https://www.theguardian.com/technology/2016/jul/29/cozy-bear-fancy-bear-russia-hack-dnc.

Thompson, Loren. "Boeing to Build Its First Offshore Plane Factory in China as Ex-Im Bank Withers." *Forbes*, 15 September 2015.

Threat Intelligence. "APT28: A Window into Russia's Cyber Espionage Operations?" 27 October 2014. https://www.fireeye.com/blog/threat-research/2014/10/apt28-a-window-into-russias-cyber-espionage-operations.html.

Tiezzi, Shannon. "China Push for an Asia-Pacific Free Trade Agreement." *The Diplomat*, 30 October 2014.

Trend Micro, Inc. "AsiaInfo to Acquire Trend Micro Chinese Subsidiary." Press release, 31 August 2015. http://newsroom.trendmicro.com/press-release/financial/asiainfo-acquire-trend-micro-chinese-subsidiary.

Trevedi, Anjani. "China, Japan Shed U.S. Treasury Holdings." *The Wall Street Journal*, 18 November 2015. http://www.wsj.com/articles/china-japan-shed-u-s-treasury-holdings-1447824480.

Tsvetkova, Maria. "Special Report: Russian Fighters, Caught in Ukraine, Cast Adrift by Moscow." Reuters, 29 May 2015. http://www.reuters.com/article/us-ukraine-crisis-captured-specialreport-idUSKBN0OE0YE20150529.

Tzu, Sun. *The Art of War*. London: Luzac, 1910.

United States–China Economic and Security Review Commission. *2010 Annual Report to Congress*. 111th Congress, 2nd Session, November 2010.

_____. *2015 Annual Report to Congress*.

_____. *2016 Annual Report to Congress*.

_____. "National Security Implications of Investments and Products from the People's Republic of China in the Telecommunications Sector." Staff report January 2011.

United States Justice Department. "U.S. Charges Russian FSB Officers and Their Criminal Conspirators for Hacking Yahoo and Millions of Email Accounts." Press release, 15 March 2017. https://www.justice.gov/usao-ndca/pr/us-charges-russian-fsb-officers-and-their-criminal-conspirators-hacking-yahoo-and.

United States Strategic Command Fact File. http://www.stratcom.mil/fact_sheets/fact_jtf_gno.html.

_____. http://www.stratcom.mil/fact_sheets/fact_jioc.html.

Urrich, Lawrence. "Chinese-Made Cars Arrive in U.S. Showrooms." *The New York Times*, 28 January 2016.

USA Today. "China Closes 13 Walmart Stores and Arrests 2 Employees." http://www.usatoday.com/money/industries/retail/story/2011-10-13/China-Wal-Mart/50751382/1.

Vozzella, Laura, and Simon Denyer. "Donor to Clinton Foundation, McAuliffe Caught Up in Chinese Cash-for-Votes Scandal." *The Washington Post*, 16 September 2016. https://www.washingtonpost.com/local/virginia-politics/clinton-foundation-mcauliffe-donor-caught-up-in-chinese-cash-for-votes-scandal/2016/09/16/bfb3b8fc-7c13-11e6-ac8e-cf8e0dd91dc7_story.html?utm_term=.47cf8561f1b7.

Washington Post. "Will China Keep Its Cyber Promises?" 21 October 2015.

Wheeler, Travis. "China's MIRVs: Separating Fact from Fiction." *The Diplomat*, 18 May 2017. http://thediplomat.com/2016/05/chinas-mirvs-separating-fact-from-fiction/.

Williams, Katie Bo. "NSA Head: DNC Hack Didn't Affect Election Outcome." *The Hill*, 21 November 2016. http://thehill.com/policy/cybersecurity/307031-nsa-head-dnc-hack-didnt-impact-election-outcome.

Wilson, Clay. "Information Operations, Electronic Warfare, and Cyberwar: Capabilities and Related Policy Issues." Congressional Research Service, 20 March 2007.

Wong, Chun Han. "China Appears to Have Built Radar Facilities on Disputed South China Sea Islands." *The Wall Street Journal*, 23 February 2016.

Worldnet Daily. "America's China Syndrome." 19 August 2003. http://www.wnd.com/?pageId=20358#ixzz1b2qADEoL.

Xecun, Murong. "Scaling China's Great Firewall." *The New York Times*, 17 August 2015.

Yan, Sophia. "U.S. Runs Out of Investor Visas Again as Chinese Flood Program." CNN, 15 April 2015. http://money.cnn.com/2015/04/15/news/economy/china-us-visa-eb5-immigrant-investor/.

Zalmay, Khalilzad, and John P. White. *The Changing Role of Information in Warfare*. Santa Monica, CA: Rand, 1999.

Zedong, Mao. *Mao Tse-Tung on Protracted War*. Peking: Foreign Languages Press, 1967.

Zetter, Kim. "Experts Are Still Divided on Whether North Korea Is Behind Sony Attack." *Wired*, 23 December 2014.

_____. "Inside the Cunning Unprecedented Hack of Ukraine's Power Grid." *Wired*, 3 March 2016. https://www.wired.com/2016/03/inside-cunning-unprecedented-hack-ukraines-power-grid/.

_____. "U.S. Considered Hacking Libya's Air Defense to Disable Radar." *Wired*, 17 October 2011. http://www.wired.com/threatlevel/2011/10/us-considered-hacking-libya/.

Index